MUSCULOSKELETAL TRAUMA

Translated from the German Series:

Krankengymnastik
Taschenlehrbuch in elf Bänden

Edited by

H. Cotta, W. Heipertz,

A. Hüter-Becker, G. Rompe

Clinical
Handbooks
in
Physical
Therapy
Management

U. Bergmann
W. Heipertz
B. Morgenroth
L. Zichner

MUSCULOSKELETAL TRAUMA

English Language Edition Edited by
Mary Anne Rinehart, M.S.
Santa Clara Valley Medical Center
San Jose, California

and

Tom Sutton, M.B.A.
Davies Medical Center
San Francisco, California

translated by
Renate King, D.V.M.

AN ASPEN PUBLICATION®
Aspen Publishers, Inc.
Rockville, Maryland
Royal Tunbridge Wells
1987

Library of Congress Cataloging-in-Publication Data

Traumatologie. English.
Musculoskeletal trauma.

(Clinical handbooks in physical therapy management)
Translation of: Traumatologie.
"An Aspen publication."
Includes bibliographies and index.
1. Wounds and injuries. 2. Physical therapy.
I. Bergmann, U. II. Rinehart, Mary Anne. III. Sutton,
Tom, M.B.A. IV. Title. V. Series. [DNLM: 1. Wounds
and Injuries. WO 700 T7775]
RD93.T74513 1987 617'.47044 87-954
ISBN: 0-87189-624-9

Library of Congress Catalog Card Number: 87-954
ISBN: 0-87189-624-9

Editorial Services: Jane Coyle

Printed in the United States of America

1 2 3 4 5

Table of Contents

Publisher's Note

Aspen Publishers is pleased to announce this series entitled *Clinical Handbooks for Physical Therapy Management*. These handbooks have been selected from a prestigious series entitled *Krankengymnastik*. Prior to the current edition, this German series was collected together as one single volume and it served as a standard text/reference for decades. Recently, the German editors, H. Cotta, W. Heipertz, A. Hüter-Becker, and G. Rompe, along with their publisher, Georg Thieme of Stuttgart, decided to issue the new edition as a series of ten smaller topical volumes to allow for greater flexibility in the use of the material.

We selected those volumes in the series that we felt were most appropriate to North American and other English language readers. Each volume we select is translated by a biomedical expert and then heavily edited for language usage. Finally, each volume is edited by American experts for content and appropriateness of coverage and terminology.

We are indebted to our experts who serve as Editors of our editions and to Susan Woll, PT, who lent her knowledge of content and language to all titles in the series.

These volumes will be distributed to readers worldwide. We hope the English language editions serve our readers as well as their German counterparts have.

Preface

The focus of *Musculoskeletal Trauma* is to identify the various musculoskeletal problems that may result from acute trauma. The book is designed to provide an understanding of the mechanisms of injury and it gives basic guidelines for the medical and physical therapy management of the patient with an acute trauma injury.

Musculoskeletal Trauma heavily emphasizes the use of PNF techniques in both the acute and rehabilitation stages of treatment. Pool therapy is often used as an adjunct to treatment, more frequently than has been our experience in the United States.

The book does not include suggested treatment for total joint replacements other than total hips, nor does it address the long-term rehabilitation of trauma patients such as those with spinal cord injuries or head trauma. Manual therapeutic techniques to increase joint range of motion are not discussed.

This book is best used as a reference in a physical therapy department or in a physical therapy educational program. The reader can obtain information about the mechanisms of injuries and treatment suggestions that may prove helpful in planning individualized treatment programs for patients.

Mary Anne Rinehart
Tom Sutton

Part I

General Principles

Chapter 1

Introduction and Description of Traumatic Injuries

L. Zichner and W. Heipertz

Trauma is an injury resulting from an external force applied to the organism, directly or indirectly. Injuries result from different mechanisms: blow, impact, projectile, temperature, radiation, and so on. The severity of the traumatic event is dependent on the consistency of the objects (size, weight, surface structure, velocity) and the resistance of the body region or tissue. Trauma care is concerned with the acute treatment and rehabilitation of injured patients. Traumatology goals are to gain an understanding of the causes and consequences of injuries and to assess the results of therapeutic intervention.

Multiple trauma is defined as simultaneous multiple injury to several areas of the body. It must be differentiated from several fractures in the same bone or multiple fractures in several bones. Multiple trauma (approximately 20 percent of hospitalized patients) has a very serious prognosis. Most frequently, the skull and chest are injured together. Thoracic and abdominal injuries are more serious but occur less frequently.

Initially, the vital functions have to be stabilized. This requires a diagnosis of life-threatening injuries to the respiratory, circulatory, and central nervous systems. After the initial procedures—intubation and artificial respiration, intravenous catheter and fluid replacement—the following further clinical evaluation and treatment may be instituted, depending on the urgency of the situation:

1. Life-saving emergency treatment, for example, in the case of spleen or liver rupture or pericardial effusion
2. Immediate selective treatment, for example, stabilizing a compound fracture
3. Treatments that can be delayed for days or weeks, for example, suture for a peripheral nerve or tendon

Stabilizing a fracture facilitates intensive care. Open fractures have priority in medical treatment over closed fractures. After stabilizing the fracture, physical therapy is indicated, even when the patient is unconscious.

Physical therapy is usually based on the diagnosis, which is further dictated by the extent of the injury and the anticipated results of surgical and medical management. Quite often a professional team is formed to provide care to a patient. In such cases, the physical therapist plays an important role in cooperation with physicians, nurses, and occupational therapists. Medical, physical, and psychological treatment is aimed toward maximum functional recovery. In patients with permanent disabilities, physical therapy aims at minimizing the continuing limitations and achieving maximal independence with remaining disabilities.

The initial medical care should encompass the immediate and future requirements of a patient. The initial care sets the stage for effective therapeutic intervention. This is particularly crucial in patients with multiple trauma. Indeed, the first procedures at the site of an accident are the deciding factors in the continuing treatment and its resultant success. In short, early intervention lays the foundation for functional treatment and a satisfactory rehabilitation.

The success of surgery is often dependent on the results of preparatory, accompanying, and concluding physical therapy. These therapeutic interventions require special instructions and planning. Other related methods, for example, occupational therapy, should be included in the intervention program. Sports and games can be very valuable techniques during the rehabilitation phase.

Physical therapy, in the field of traumatology, requires basic anatomic and physiologic knowledge, an understanding of trauma types and mechanisms, and the implementation of medical treatments and goals in combination with therapeutic techniques. This is the focus of this book. Areas of lesser importance to physical therapy (e.g., shock, first aid, internal injuries, paralysis) are noted only briefly, and the reader is referred to other publications for additional information on those subjects. Specifically, we will be concerned with acute injuries to the musculoskeletal system, complications resulting from the injuries, and the treatment of these injuries and their complications. Medical treatment is discussed to the extent necessary to understand the therapeutic recommendations, directions, and precautions. The types of injuries in which physical therapy is most needed and beneficial are given particular attention.

Whenever possible, functional treatment methods for trauma patients are emphasized. In some cases, however, as in the case of a fractured calcaneus or humeral head, therapy to restore function cannot be applied unless stability of the fracture site is secured by surgical fixation. As soon as musculoskeletal stability is achieved, physical therapy works toward the restitution of functional ability.

WOUNDS

Wounds and Wound Healing

Skin injuries, generally called wounds, are common. They can be caused mechanically, thermally, or chemically. The usual lacerations of skin or mucosa have a typical appearance related to the type of trauma: cuts have even edges; blows cause contusion with smooth wound edges; stab wounds usually penetrate deeply into the tissue; tears have irregular edges; bullet wounds cause deep and extensive destruction; bite wounds can be complicated by secondary infection (Figure 1–1).

In the healing of a primary wound, the defect is covered by supporting collagen tissue, which is then replaced by epithelial growth. Wounds can heal by themselves (primary healing) if the edges are close enough together and can be bridged by newly formed tissue. In more extensive wounds, the edges have to be approximated surgically by sutures or skin grafts. Wound healing that occurs as a result of intervening surgery is termed secondary healing. Granulation tissue is formed, and the contracted wound is gradually filled in and finally covered with epithelial tissue. Secondary healing often results in unfavorable and extensive scars with adhesions to the subcutaneous tissue.

In surgical wound care, one tries to create conditions for primary healing by debriding the wound edges until healthy viable tissue appears approximating the peripheral areas to avoid too much tension of the sutures, using careful suture techniques, and applying a sterile dressing. The site of the injury is elevated to decrease edema, and tetanus prophylaxis is provided as indicated.

Wound adhesions and contractures, especially those limiting range of motion, are treated with physical therapy to increase tissue elasticity and joint range of

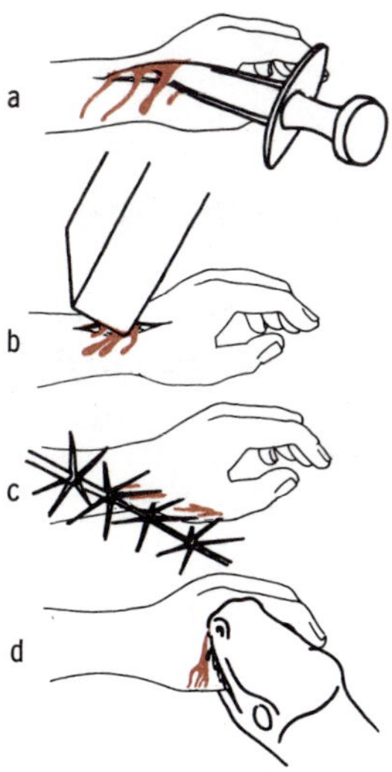

Figure 1–1 Common wound types: a. cut and stab wound, b. contusion, c. laceration, d. bite wound

Source: Schlosser and Kuner, 1980

motion. Special ointments combined with massage improve the scar flexibility. In the case of prolonged bed rest required by extensive injuries, prophylactic treatment against thrombosis and embolism is mandatory.

Structural Changes In Scars and Connective Tissue

Larger wounds that involve deeper tissue structures result in extensive and visible scar formation, for example, in compound fractures, open joint injuries, restorative surgery, and amputations. Such wounds might include connective tissue damage that requires special attention. In these cases, the connective tissue has a supporting and gliding function and also serves as nutritive tissue. Pathologic alteration of connective tissue plays a major part in immobilized and/or contracted joints and in Sudeck's atrophy.

The healing process of more extensive skin ulcerations and injuries occurs mainly by epithelial repair from the periphery to the center. This scar formation can be accelerated by massaging, thereby avoiding adhesions and improving local hyperemia. Scarring adhesions are especially problematic in amputation stumps. The therapeutic management against scar contraction has to be initiated early in order to utilize the fairly good elasticity of connective tissue.

Sometimes the regenerating connective tissue grows into hypertrophic dimensions, which interferes with the healing process of the adjacent tissue. The overgrown connective tissue has the tendency to shrink and this requires stretching and softening techniques (according to Kohlrausch) that not only treat the contracture but also promote the absorption of hypertrophic connective tissue. Stretching has to be conducted gently and repeatedly for fairly long periods, since any abrupt stimulus will enhance the risk of more scar formation.

Permanent edematous formation is very damaging to connective tissue. Following the injury, cavities with serous lining and lymphatic effusions can be observed in the skin and may require surgical intervention. Lymphatic drainage can be seriously hampered by traumatic separation of the skin from the underlying fascia and connective tissue.

Acute subcutaneous hematomas are treated with ice packs, pressure bandages, and temporary immobilization. With older hematomas, heat and exercise can be utilized.

INJURIES OF THE MUSCLE AND TENDON

The effects of therapy in the case of injuries of the muscle and tendon are variable. Muscle tissue is highly vascularized and therefore provides favorable healing conditions. Tendons do not have blood vessels and are bradytrophic tissue. Like tendons, fasciae and menisci because of their low metabolism do not have much regenerative capability. The nutrition of tendons is supplied in equal parts by the adjacent muscle, through the bony insertion, and by diffusion from the tendon sheath.

Open Injuries of the Muscle

Open wounds that penetrate skin and subcutaneous tissue create an increased risk of infection by contamination from the foreign bodies causing the wound, such as metal, wood, or bullets. Acutely disrupted tissue or deep cavities are ideal environments for anaerobic-growing tetanus or clostridial bacilli and require careful corrective surgery. To achieve undisturbed healing and complete functional restitution the edges of the wound must be trimmed and shaped, necrotic tissue must be removed, and muscle stumps must be adapted.

Muscle tissue does not heal by forming new contractile fibrils. The healing occurs through connective tissue replacement, which does not cause a functional loss. Extensive muscle damage is accompanied by general clinical symptoms, such as hypovolemic shock, or with a crush injury by renal loss of myoglobin, which can cause ischemic muscle necrosis and uremia.

Muscle Rupture and Strain

Muscle ruptures are caused mainly by the body's own mechanical forces; very seldom are they caused by an external force such as muscle contusion by an underlying bone. Because of the disproportion between the applied force and tissue stability, particularly when muscle coordination is lacking, a muscle ruptures partially or completely. Spontaneous ruptures can occur after a "wrong" movement in athletic activities, or even during daily routine activities that do not involve extreme force. The latter ruptures are caused by inflammatory or degenerative alteration of the muscle.

The mildest muscle injury is a muscle strain. This is caused by microtears in single muscle fibrils and does not need special attention. However, in case of decreased blood circulation or muscle tension, physical therapy should be used to improve the circulation and relieve muscle spasm.

More extensive acute muscle ruptures can be detected through palpation as painful impressions in the tensed muscles. These require immobilization followed by bandaging, taping, or splinting. Complete muscle ruptures have to be corrected surgically, then immobilized for four to six weeks, after which physical therapy should be carefully applied. The injury itself, or its unprofessional treatment, can lead to myositis ossificans. The main reason for this is reduced muscular nutrition caused by damaged blood vessels. A myositis caused by a damaged nerve supply is referred to as a neurotically circumscribed ossifying myositis.

Open Injuries of the Tendon

Tendons can be injured by deep cuts, stab and bullet wounds, contusions, or by burns. The interrupted continuity becomes obvious in a loss of function. The therapy depends on the severity of the lesion, whether there is partial or complete separation, and its location. A complete injury requires a surgical repair. Intensive physical therapy in combination with occupational therapy is necessary before and after plastic tendon repair.

Bacterial tendosynovitis can develop even though there is no open injury. Bacterial invasion causes a severe purulent inflammation that progresses within the tendon sheaths. The so-called V-inflammation is especially serious; this can develop as a result of a connection between the sheaths of the flexor tendons of the thumb and the last finger. Immobilization for up to six weeks is necessary after tendon surgery, not only to protect the sutures but to avoid the risk of progressive infection. Immobilization may be necessary in a stab injury, such as that caused by a penetrating rusty nail. Mobilization might be possible in certain cases, for example, in the case of sutured flexor tendons.

Tendon Rupture and Strain

Closed tendon injuries range from so-called strains to complete ruptures. Complete ruptures and small lacerations are observed most frequently. Ruptures of major tendons are often caused by an underlying tissue degeneration. Less frequently they are the result of uncoordinated forceful movements of the body. In many cases, the clinician is dealing with a combination of causes, such as premature or age-related wear and permanent stress or deformation within the tendon sheath.

Because of the considerable diameter of tendons attached to major muscles, a healthy tendon can endure a pronounced pulling of the attached muscle. The transmission of mechanical force by a tendon is related to the lack of elasticity. An unaffected tendon will most likely tear at the junction of the tendon and muscle or be completely separated at the bony insertion point, that is, separation of the extensor tendon at the distal phalanx of the finger or the tendon at the olecranon. Ruptures of the Achilles tendon may be caused by excessive stress to the tendon. This may be due to an abrupt muscular stress or the imposition of a blunt force against a tensed tendon. Acute Achilles tendon ruptures without preceding tissue degeneration are seen most often in athletic injuries.

A rupture of the quadriceps tendon is caused mainly by an age-related process during the second half of the life span. The same process is involved in ruptured tendons of the biceps, supraspinatus, and the extensors of the thumb and fingers. In a traumatic rupture of the supraspinatus tendon, a bony lamella is usually torn off, together with the tendinous insertion.

Generally, an acute tendon rupture is treated surgically. However, a ruptured biceps tendon may be managed without surgery. Chronic tendon ruptures in upper extremities are frequently conservatively treated to achieve optimal function, while plastic surgery is performed on the lower extremity to reestablish a physiological function.

INJURIES OF THE PERIPHERAL NERVE

Injury to a nerve is obvious immediately as a result of the loss of motor and sensory function in the area supplied by the damaged nerve. Secondary symptoms are trophic alterations of the skin, blood vessels, muscles, and bone. Muscle atrophy may lead to joint deformities. Nerves are involved in approximately 5 percent of all traumatic injuries.

Peripheral nerve injuries have different healing tendencies, depending on the severity and type of trauma—mechanical, electric, thermal, or chemical—or the direct result of a loss of blood supply. Peripheral nerve injuries are classified as:

- neurapraxia (demyelinization with remaining neural axon)
- axonotmesis (interruption of the axon)
- neurotmesis (complete disruption of the nerve)

Spontaneous regeneration can be expected in neurapraxia and is possible in axonotmesis. In neurotmesis, regeneration is usually incomplete, despite sutures. Restoration during nerve regeneration proceeds in the following sequence: sweat gland functioning, perception of cold and warm temperatures, sensitivity to touch, and finally motor function. After approximation of the neurolemma sheath, the regeneration advances approximately 1 mm per day.

Open Injuries of the Nerve

Nerve injuries in a clean wound are repaired with sutures. Otherwise, the treatment involves surgery after the wound is clean. Even later on (six months or longer), a suture repair is possible; however, in such cases the chances of healing are not good, because of the atrophy of the distal nerve sheaths, the motor end plates, and the sensory organs.

In a complete interruption, the nerve stumps have to be approximated by a suture. In more extensive defects that cannot be closed without tension, a cable transplant is necessary to bridge the gap. Repaired nerves have to be immobolized for several weeks to protect the regeneration phase. Afterward, careful physical therapy begins with passive exercises, followed immediately by active exercises using functional splints.

Closed Injuries of the Nerve

Closed nerve injuries caused by pressure or distortion are evidenced initially by strain in movement. In such cases, exercises and electrotherapy have been successful. Electrotherapy does not accelerate the nerve regeneration; it rather retards the atrophy of the denervated muscle fibers and thus maintains a functional organ for the regenerated nerve.

If continuous clinical observation does not reveal signs of regeneration, the nerve might benefit from surgical revision. This is particularly relevant in cases of increasing neurologic deficits following local trauma from a fracture or joint subluxation. In surgical revision, the nerve is exposed and decompressed, an especially important procedure in the case of pressure on the peroneal, radial, and ulnar nerves.

A disuse atrophy of the thigh muscles during and after knee joint injuries can lead to an erroneous diagnosis of a paresis of the quadriceps femoris muscle due to a damaged femoral nerve. The atrophy resulting from inactivity involves not only the knee extensor muscles, innervated by the femoral nerve, but also the hip flexor and adductor muscles. However, there is no loss of sensitivity, as in the case of a lesion of a peripheral nerve. If necessary, the correct diagnosis can be achieved by electromyography. In the diagnosis and treatment of nerve injuries one has also to consider anatomically abnormal innervations of muscles and dermatomal areas supported by neighboring nerves.

Untreated nerve injuries can result in joint contractures. Typical symptoms of such injuries are related to the effect of innervated antagonist muscles and noninnervated agonist muscles. Examples are a "clawhand" in loss of the ulnar nerve, a wristdrop in loss of the radial nerve, a "monkey" hand in loss of the median-ulnar nerve, an equinus in loss of the sciatic or peroneal nerve. During the first year after the injury, the muscle atrophy develops within the first few weeks and then progresses into fiber degeneration if peripheral nerve regeneration does not occur.

The vasomotor trophic disturbances following nerve injuries consist of, first, an increase of blood supply to the afflicted area during the first few weeks, then a diminished blood supply, and finally atrophy. Additional trophic symptoms include hyperkeratosis, trophic skin ulcers with little healing tendency, disturbed growth of hair and nails, decreased pigmentation, and sweat secretion.

INJURIES OF THE JOINT

Joint injuries can include contusions, sprains, ruptured ligaments, and subluxations. Symptoms of joint injuries include pain, limited range of motion, edema, hyperemia, and hematoma. Increased capillary permeability results in an articular effusion, which can cause an overstretching of the joint capsule. Accompanying fractures and ligament ruptures can be verified or excluded by radiograms.

Contusion

Contusions are caused by direct blunt trauma, such as a blow, push, or fall against a hard obstacle, or by pinching. Tissue damage can include damage to the skin, subcutaneous tissue, muscles, fasciae and tendons, joint capsules and ligaments, or cartilage and bone. Contusions of the joint cartilage, involving cartilaginous areas with underlying spongy bone and cartilage that is completely torn from the surrounding tissue, can create serious joint problems. Uneven intra-articular surfaces produce premature wear and breakdown of the cartilage, which may be further exacerbated by freely floating loose bodies in the joint space. Ruptured vessels cause hemorrhage. Extensive subfascial tissue damage and contraindicated therapy, especially massage, can create calcifications in the muscle, myositis ossificans. The most frequently involved areas in joint injuries are the foot, hand, and finger, followed by the shoulder, elbow, and knee, and finally the spine and hip.

Bone contusions, mainly on the shin bone and skull, can cause periosteal irritation and inflammation. Articular bone and cartilage are affected mainly by indirect or transmitted contusions.

In a contusion, the pain is initiated by pressure and motion, but it can also occur spontaneously. The edema is caused by excessive fluid or blood in the tissue. A subcutaneous hematoma produces discoloration of the skin.

Contusions are treated by rest, elevation, cold packs, and pressure bandages. Muscle spasms are loosened by gentle exercise and local ice application.

Sprain and Ruptured Ligament

Sprains occur as a result of a motion that exceeds the physiologic joint range or a force transmitted indirectly in an abnormal direction. A minor sprain is an elastic and reversible stretching that can heal without complications. A more extensive sprain is characterized by damaged muscle fibrils and ligaments that are partially ruptured but have intact continuity.

The pain in a sprain occurs from pressure, stretching, and motion of joint structures. Joint range of motion is limited by pain with the formation of edema and hematoma. Frequently, articular ligaments as well as muscles and tendons are involved. Sometimes the peripheral nerves are strained. Sprains of the knee joint (medially) and the ankle (laterally) are the most frequent diagnoses.

For the first few days, the treatment of a sprain is the same as that for a contusion. In the case of persisting complaints, immobilization is indicated for two to six weeks.

Severe sprains are characterized by torn ligaments and joint capsules, with interrupted continuity of one or several ligaments, or by ligaments completely separated from their insertion at the bone. The resistance of ligaments is weakest at the point of their bony insertion, which results in frequent lacerations at that point. Ligament ruptures occur in the following order of frequency: (1) proximal tarsal joint (lateral syndesmosis), (2) knee joint (medial and cruciate ligaments), (3) finger joints (metacarpals I, II, V), (4) elbow joints, (5) shoulder joints. The capsule seldom is torn across the joint cavity.

Although the extent of a ligament injury in a severe sprain may be equal to that of a subluxation, the initial symptoms are like those of a minor sprain. However, the minor sprains still require care, since insufficient therapy would result in an instable joint. Medical management for ruptured ligaments should be based on the diagnosis of an instable joint. The ligament rupture should be confirmed by manual tests and radiograms. A standardized radiographic evaluation is mandatory to diagnose or exclude bony lesions—including ligament ruptures at the bony insertion.

Severe sprains with partial ruptures of the capsule and torn ligaments require immediate surgery with approximation sutures. Surgery is especially recommended for affected knee joints, the tarsus, and the proximal joints of the thumb (see Part II). Afterward, splints or walking casts are applied for four to six weeks. For an ankle sprain a high top shoe might be used as an alternative method of stabilization.

Nonsurgical treatment consists of immobilization in a splint, followed by physical therapy. But frequently the result of this is an incomplete restitution of ligament stability. A persistent instability requires an operation to achieve passive and active joint stability, especially in young patients or active athletes.

In diagnosing and treating sprains, one has to consider the involvement of articulations close to the joint, such as the radioulnar syndesmosis in wrist sprains or the tibiofibular syndesmosis in ankle sprains. The clinical symptoms of frequent strains of the shoulder are often persistent and painful. This most commonly involves the muscles and tendons of the rotator cuff. The shoulder joint is partly stiff as a result of immobilization and/or lack of exercise, especially if there were prior lesions, such as a degenerative process of the cervical spine. Therefore physical therapy for a post-traumatic stiff shoulder is often lengthy; occasionally, it requires mobilization under anesthesia, followed by an abduction (airplane) splint and physical therapy.

Sprains of the knee joint by an indirect or twisting force can cause meniscus lesions. However, if there is only a little or ineffective force and the meniscus lesions are mainly caused by degeneration, it is preferable to refer to the condition as meniscus detachment. In any case, the damaged meniscus must be removed surgically (see also the section on knee joints).

Subluxation

A subluxation is a dislocation of the joint-forming bones against each other. It includes a partially or completely torn joint capsule and ligaments resulting from direct or indirect trauma. Subluxations are classified in terms of their origin:

- traumatic subluxation, caused by a strong and sudden impact during excessive flexion, extension, abduction, or adduction
- habitual or recurrent subluxation, caused by a physiological load on an unstable joint
- pathologic subluxation, caused by a gradually increasing physiological load on pathologically damaged joint components
- congenital subluxation, caused by joint dysplasia exposed to a slowly increasing physiological load

The joints most frequently involved are:

- traumatic: shoulders (50 percent), elbows (20 percent), finger and toe joints, proximal tarsal joints, acromioclavicular joints, hips, carpal joints, and knee joints
- nontraumatic: shoulders, femoropatellar joints, proximal phalangeal thumb joints

The general symptoms of a subluxation are pain and swelling around the joint and limited range of motion. The diagnosis can be confirmed by deformation and malposition of the joint components.

An especially strong force is necessary to sublux a deeply anchored spherical joint, such as the femoral head gliding within the bony acetabular socket. This type of subluxation is frequently complicated by split bone fragments. In an existing tendency to sublux as a result of a dysplastic joint, only a little force is necessary, as in nontraumatic subluxation of the shoulder and patella. The correct diagnosis and possible verification of accompanying bone lesions is based on manual tests and radiographs. Control radiographs are indicated after a reposition. Possible ligament ruptures will also show up on radiographs.

In a subluxation, the major complications are lesions of neighboring nerves and a decreased blood flow in the limb distal to the injury. Pressure on the injured joint structures produces local necroses of the intra-articular cartilage, or even splitting of cartilage and bone compound. In a subluxation, there is always a risk of a capsular shrinkage, since the injured ligaments and capsules undergo a scarring contraction. Hematomas in the intra-articular space or the surrounding tissue may further limit the range of motion during the organization and resorption phase. The shoulder joint in particular tends to have a limited range of motion, especially after long immobilization in an abduction (airplane) splint. Late complications caused by subluxations include periarticular calcifications or ossifications.

In a subluxation, the primary care is a suitable position, possibly in a splint, during the initial transportation, then a reduction of the joint, preferably under anesthesia. An immediate reduction is advisable to avoid, or at least reduce, complications. This is especially recommended in the case of neurologic or circulatory disturbances.

Surgery is indicated in cases of unsuccessful repositioning, simultaneous injuries of capsules, ligaments, and major vessels, or for the purpose of decompressing a nerve. Further surgical manipulations are needed to reposition and fixate bone fragments in an accompanying fracture or in an open reduction of joint components.

After the initial medical management, the healing of capsule and ligaments has to be protected by an immobilization period of two to six weeks, depending on the extent of the lesion. The joint is positioned to prevent adhesions and scar formation from interfering with a physiological function. Afterward, physical therapy concentrates on regaining joint movement and function.

In a nontraumatic subluxation, a reduction is usually successful and does not require anesthesia. The immobilization to relieve the pain is applied only for four to six days. A tendency to sublux caused by trauma but based on a congenital disposition must also be corrected surgically.

Shoulder subluxations are particularly serious. Not only the shoulder joint itself is affected, but also the acromioclavicular joint and the sternoclavicular joint. Surgery and temporary fixation might be required together with ligament sutures.

Subluxating shoulder joints in adolescents, who are able to voluntarily elevate the humeral head partially out of the socket and let it snap back with a distinct click, are presumed to be caused by a joint instability or faulty innervation. In such cases, the patients follow a program of shoulder strengthening exercises before they undergo surgery to correct the dysplasia.

Myositis ossificans is observed frequently after a subluxation of the elbow joint. Therefore, appropriate physical therapy, avoiding massages, must be applied for approximately two weeks following surgical reduction and immobilization. It is advisable for prophylactic reasons to treat with diphosphates.

Subluxations of finger joints are usually caused by falling on the hand or catching a ball. Occasionally, they are complicated by small split-off bone fragments. These subluxations can often be corrected without, or with only local, anesthesia. Accompanying tendon injuries have to be treated surgically.

FRACTURES: GENERAL PRINCIPLES

A fracture is a disruption of the continuity of bone. It is usually caused by a force that exceeds the elasticity of the bone. The force can be direct or indirect. In a few cases, fractures are caused not by an external force but an internal force, as, for example, with an endogenous or stress fracture in athletes.

In children and adolescents with incomplete bone growth, the periosteum frequently remains intact after a fracture. This is called a ''greenstick'' fracture.

Fractures that are characterized by crevices but without disruption to the bone continuity are called fissures.

Spontaneous or pathological fractures occur without adequate trauma. They are based on pathological processes of the skeleton that, because of a minor trauma, decrease the bone stability and lead to a disruption of the continuity. Examples of basic benign processes are generalized osteoporosis and local benign tumors or tumorlike lesions. The most frequent causes of pathological fractures are primary and secondary malignant tumors. The most common of such fractures are metastases originating from breast, kidney, or thyroid tumors.

Traumatic fractures are differentiated from fatigue fractures caused by chronic overload. Traumatic fractures originate from extreme malposition such as traumatic coxa vara or from activity such as marching fracture.

Structures adjacent to the fracture site, especially vessels and nerves but also muscles and tendons, can also be affected.

An open fracture is characterized by skin penetration and exposure of the bone fragments. This kind of fracture, also called a ''complicated'' fracture, has an increased risk of infection. Such fractures are classified as:

- first degree open fracture: small penetration from the inside without major trauma to the surrounding tissue
- second degree open fracture: extensive skin penetration from the inside or outside with little trauma to the surrounding tissue
- third degree open fracture: extensive penetration of all tissue structures down to the fracture

The systemic effects of a long bone fracture are mainly related to circulation. A fracture of the femur or pelvis can cause a hemorrhagic shock because of extensive blood loss. Another dangerous complication is the formation of fat emboli. This occurs most frequently after a fracture of long bones, after extensive tissue damage in severe contusions, or in burns. Resuscitation procedures can also cause emboli.

Classification of Fractures

Fractures, classified by the direction in which the bone is disrupted, are described as transverse, oblique, longitudinal, comminuted, and spiral (see Figure 1–2). Depending on the mechanism, fractures are caused either directly by an external force or indirectly by internal stress. A strong, quick impact produces a flat fracture line, whereas a slow impact leaves a steep line. Additional bending and/or contusion can produce wedge-shaped fragments (two wedge-shaped fragments are called a "butterfly" fracture). Spongy bones (vertebrae, calcaneus, tibial head) are subject primarily to compression fractures from severe deformation. A compression fracture is caused by a sudden strong pressure, affecting mainly bones adjacent to joints and vertebrae. The majority of fractures in long bones are caused by lateral pressure. A spiral fracture is caused by a pressure load in rotation.

The fracture type is often referred to in terms of the fracture site, for example, a medial or lateral femoral neck fracture. Especially in adults fractures are described by the kind and location (see Figure 1–3):

- lateral dislocation to the side
- longitudinal dislocation
- dislocation with deviated axis
- dislocation with rotation

In children, fractures close to a joint can be problematic. A sudden impact on the open epiphyseal plate in a growing bone can interfere with the growth of the bone.

According to Aitken, the following epiphyseal injuries or fractures can be distinguished (see Figure 1–4):

- Aitken O: uncomplicated epiphysiolysis separation with or without dislocation
- Aitken I: epiphysiolysis with metaphyseal fragment

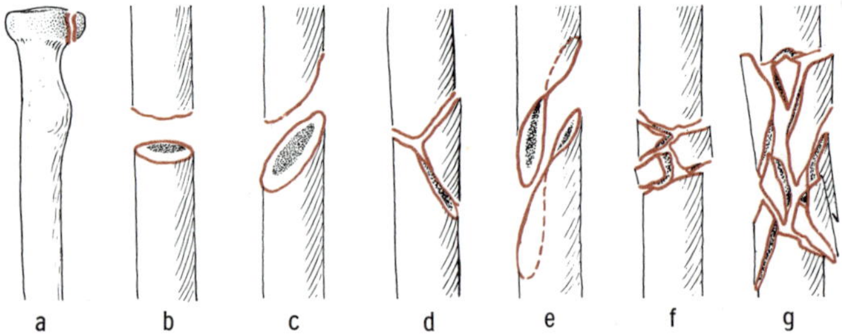

Figure 1–2 Fracture types: a. longitudinal fracture ("chisel" fracture), b. transverse fracture, c. oblique fracture, d. compound fracture, e. spiral fracture, f and g. comminuted fractures with several (f) and numerous (g) fragments

Source: Schlosser and Kuner, 1980

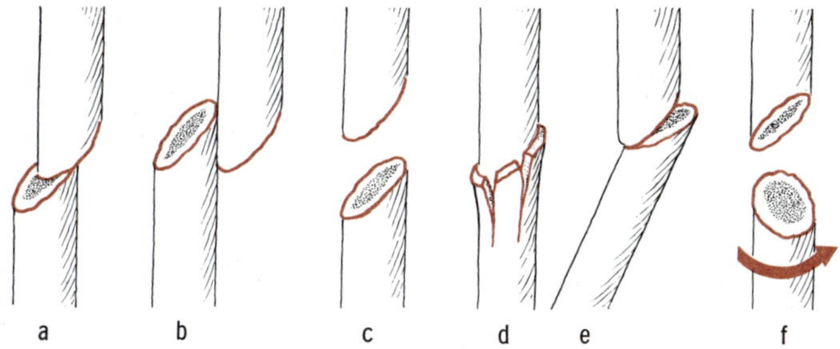

Figure 1–3 Dislocation of the fragments: a. lateral, b-d. longitudinal: b. shortening, c. elongation, d. compression, e. deviated axis, f. rotation

Source: Schlosser and Kuner, 1980

- Aitken II: fracture across the epiphyseal plate (epiphyseal fragment)
- Aitken III: fracture of the epiphyseal plate with epi- and metaphyseal fragments

A special type of injury is the epiphyseal compression with microfractures of the epiphyseal plate and spongy bone.

Clinical Symptoms

Reliable symptoms of a new fracture are bony deformation (change in shape and location), abnormal motility, a grinding of bony fragments against each other

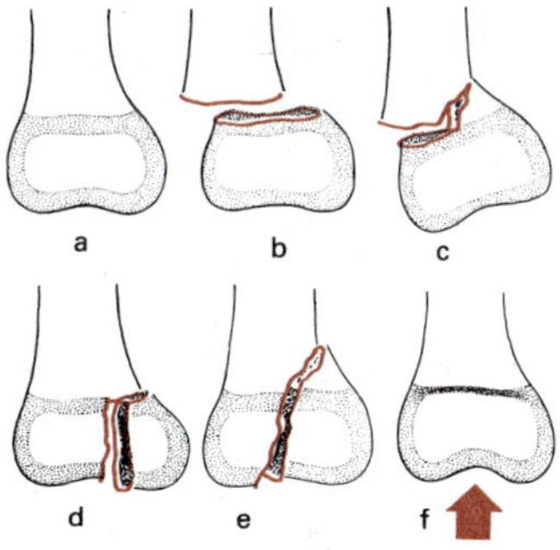

Figure 1–4 Classification of the epiphyseal injuries: a. normal configuration, b. uncomplicated epiphysiolysis (0), c. epiphysiolysis with metaphyseal wedge (I), d. epiphysiolysis with epiphyseal fragment (II), e. epi- and metaphyseal fracture (III), f. epiphyseal compression

Source: Schlosser and Kuner, 1980

(crepitation), and, in an open wound, visible bone fragments. Each of these symptoms contributes to an accurate diagnosis. Unreliable symptoms are edema (especially caused by hematoma), pain, and loss of function, since soft tissue injuries may show the same symptoms. Abnormal motility and crepitation do not occur in incomplete fractures when soft tissue is squeezed between bone fragments, and in dislocations.

Attempts to determine the existence of these symptoms can be painful. The radiographic diagnosis of a fracture is less painful and the most reliable diagnostic method. Pictures in two planes are necessary. Repeated radiographs can confirm a tentative diagnosis and also provide information on the healing process. Frequently, fissures show up in the radiograph only after ten days following the injury; also at that point, the hairlike crevice is more visible because of the onset of bone reabsorption.

Malalignment might be caused by an axis deviation or a curvature in an anterior or posterior direction. Bone fragments can shift to an extent that may be more or less than the shaft diameter. Dislocations at the fracture site can result in shortening, less often in elongation or rotation. Rotation is often not diagnosed to its full extent and, in such cases, can heal in an abnormal rotation.

Fracture Healing

Undisturbed Fracture Healing

Bone tissue is capable of regeneration. The condition for restitution of bone continuity is a bridging of the gap. As with other tissues, primary and secondary healing is possible. If the fragments are too close together to allow vascular growth from the outside (periosteum) or the inside (endosteum), the healing process originates from the vessels of the Haversian canals. This type of healing, without visible callus formation, is called primary healing. If there is a gap between the fragments (see Figure 1–5), into which endosteal or periosteal vessels can grow, the bone healing starts in the hematoma between the fragments and forms a fibrous callus. A fixation callus from an irregular netlike mesh is built by the transition of connective tissue cells into cartilage and bone-forming cells. The callus differentiates its structures in a longitudinal direction under mechanical stress. Stress causes electrical fields to be set up, which in turn cause new callus to form. After reaching a basic stability, the initially hypertrophic endosteal and periosteal callus is reduced to almost normal dimensions. The necessary conditions for this process are (1) sufficient blood supply to the bone and its surrounding tissue, (2) correct immobilization with sufficient time, and (3) clean and undisturbed conditions (no infections, cytostatic drugs, and so on).

Disturbed Fracture Healing

Disturbances occur in the following circumstances:

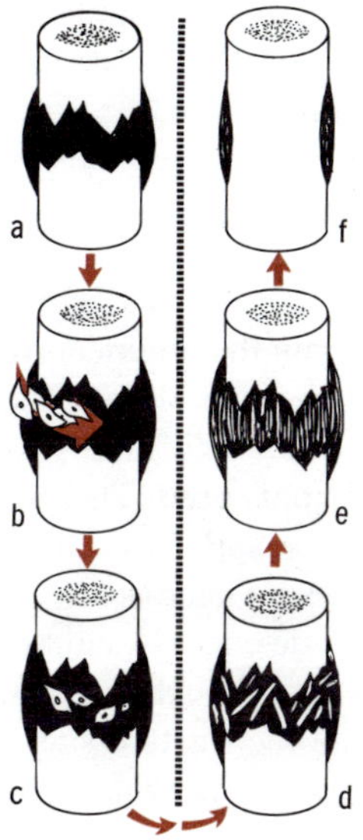

Figure 1–5 Secondary fracture healing with callus formation: a. fracture hematoma, b. invasion of connective tissue cells, c. transformation into osteoblasts, d. irregular netlike bone trabeculae (fixation callus), e. differentiated arrangement of the bone structure, f. healing by longitudinally directed trabeculae

- insufficient blood circulation through the bone and its surrounding tissue, preexisting vascular disease, and traumatically decreased circulation and its related complications, such as necrotic fragments
- insufficient immobilization (mechanical instability) during the healing phase
- local defects (lack of bone substrate, interposed muscle tissue)
- local infection (after open fractures or osteomyelitis)
- as a result of some medications (cortisone, cytostatics, anticoagulants)

If the blood supply is intact but the fracture is unstable, the periosteal callus overgrows without uniting the fragments. The result is a hypertrophic pseudarthrosis. However, if decreased blood circulation, infection, or the application of drugs interferes with the callus formation, an atrophic pseudarthrosis will occur. The fracture is bridged by connective tissue. Both pseudarthroses are characterized by weightbearing pain, instability, hyperthermia, and limited function.

Treatment of Fractures

The treatment of a fracture is aimed at the functional restitution of the affected part of the body. A complete recovery is usually achieved by the restitution of the original shape, avoiding malpositions and joint stiffness.

Basically, there are two types of surgical management: no internal fixation and fixation. Both types employ the fundamental principles of fracture treatment: (1) reposition, (2) retention (or immobilization), (3) rehabilitation.

Treatment Without Surgical Fixation

The reposition of the fragments is done under general or local anesthesia, which both eliminates the pain and provides muscle relaxation to permit the fragments to be adjusted precisely. The procedure requires that the fragments be shifted and moved with lateral pressure into the correct position. The immobilization procedure is dependent on the fracture type. Casts are most commonly used. The cast must encompass both joints adjacent to the fracture site. It may be split down to the skin to avoid post-traumatic or postoperative edema and pressure necroses.

The immobilization can be extended if a cast does not provide enough stability and/or the fragments cannot be reduced sufficiently. In this case, steady traction along the axis eliminates shortening and malpositioning, and the fracture site is protected by the surrounding soft tissue. As soon as the fracture is bridged by connective tissue, a cast is normally applied.

Rehabilitation procedures are initiated during and after the fracture consolidation. Even during the immobilization phase physical therapy can have beneficial effects, both for the fractured body area and general body systems. Preventive physical therapy is begun during the acute phase to avoid venous thrombosis. Older and weaker patients need circulatory support. Complications—such as pneumonia, pressure-related ulcers of the skin, and urinary infections (due to retrograde pressure, especially from elevated legs)—are counteracted by appropriate treatment. As soon as sufficient stability is established, an exercise therapy with increased weightbearing is initiated.

In the nonsurgical treatment process involving a considerable immobilization period, inactivity of the injured limb must be avoided. Otherwise local complications occur, including a decreased blood supply, muscle and bone atrophy, and stiff joints. More general results are decreased mobility and a limitation of vital functions. Thus, the maintenance of sufficient blood circulation is vital. Such maintenance may be endangered locally if new fractures are placed in a spiral cast and the blood drainage from and the blood flow into the fracture site are reduced by a resulting hematoma and soft-tissue edema. Nerve lesions and muscle and soft tissue necroses can then develop. Further complications due to insufficient vascular supply are calcifications, contractures, and even amputations.

Circulatory deficits while in a cast can be recognized by a bluish discoloration of the skin, soft-tissue swelling, pale distal limbs, disturbed sensitivity and motility, and pain. If initial elevation of the limb or medication for a post-traumatic edema is not successful, the cast has to be loosened immediately or removed. In open injuries, a decreased blood supply can favor the growth of anaerobic organisms, such as anthrax and tetanus.

The advantages in using a cast are that it permits early mobilization, especially in walking casts, and the fact that the fracture remains covered. The disadvantages are that immobilization is not complete and there are risks of "fracture disease" and Sudeck's atrophy.

Treatment with Surgical Fixation

In surgical treatment with fixation, the bony fragments are aligned precisely and held with a stable internal fixation device; this permits immediate postoperative physical therapy and prevents circulatory disturbances, muscle atrophy, and joint stiffening. The advantages of surgical treatment are that it permits postoperative active exercise therapy (prevention of the "fracture disease"), exact reposition, and care of the polytraumatized patient. The disadvantages are the risk of infection (1 to 2 percent) and of technical errors in surgical indication and performance. One of two methods may be employed: stable osteosynthesis or adaptive osteosynthesis.

Osteosynthesis

The main goals of a stable osteosynthesis are normal functioning of joint and muscle with a fracture site in a neutral anatomical position and a sufficiently protected healing phase. The following techniques are used:

- Medullary rod. A transverse or short oblique fracture is stabilized by using a metal rod to stabilize the two ends of the fractures, utilizing the "pipe within a pipe" principle. With this technique, even weightbearing stability can be achieved. Intramedullary pinning has the advantage of enabling movement of joints distal and proximal to the fracture site (see Figure 1–6).
- Plate. A metal plate is used when a medullary rod is inappropriate, for example, in fractures of the upper extremity and in the metaphyseal area. Compression plates are used in transverse and short oblique fractures; neutralization plates are used in comminuted fractures; and buttress plates are used in fractures of spongy bones (see Figure 1–7).
- External fixation. A two- or three-dimensional device transmits the fixating force to the fragments through percutaneously implanted nails and screws. External fixation is used in comminuted and compound fractures with or without infection (see Figure 1–8).
- Osteosyntheses, using wires, rings, or screws, are also treatment processes leading to a stable osteosynthesis (see Figure 1–9). These techniques are used

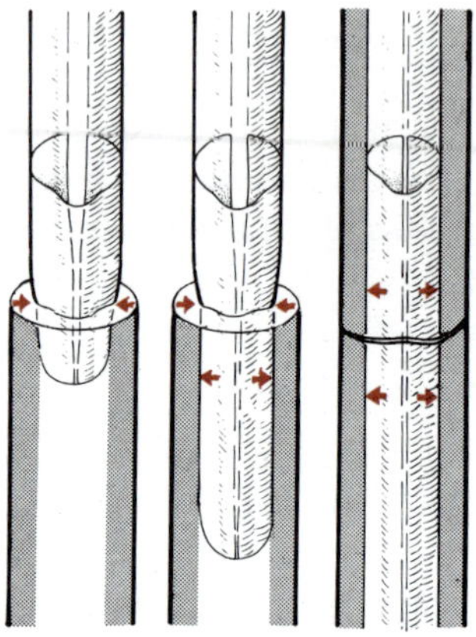

Figure 1–6 The principle of intramedullary pinning ("pipe within a pipe" system)

Source: Schlosser and Kuner, 1980

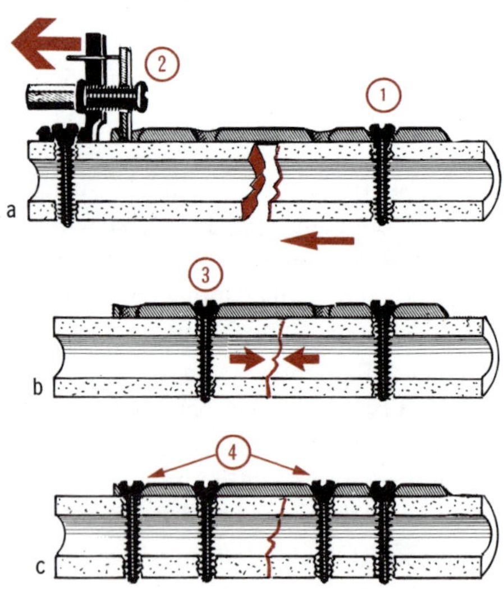

Figure 1–7 Osteosynthesis with a compression plate: a. fixation of the plate with a cortical screw on one side (1), tightened with a special device (2); b. after adaption under pressure, the fragments are aligned and the opposite side is fixed (3); c. the remaining screws are tightened (4)

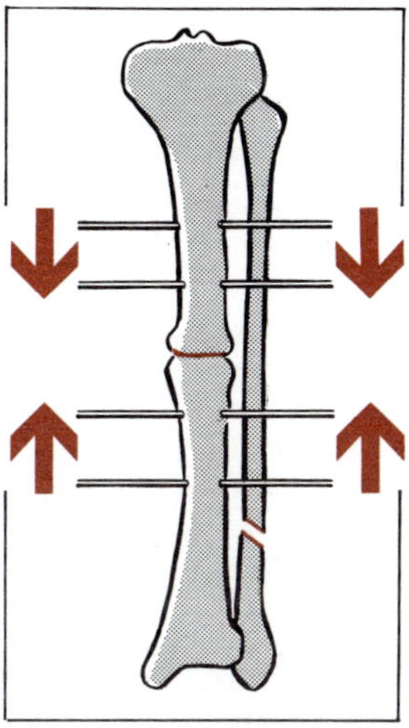

Figure 1–8 Pressure is transmitted to the fragments by two- and three-dimensional struts through Steinmann pins implanted in the proximal and distal fragment

to repair displaced fractures in children or, in some cases, to secure a realignment in adults (see Figure 1–10).

In certain types of fractures, a combination of surgery and temporary external immobilization (cast for days or weeks) is indicated, especially when the fracture is accompanied by ligament ruptures. In some fractures in old patients and in severe comminuted joint fractures, a primary endoprosthetic replacement might be required. Osteosynthesis creates optimal conditions for a completely restored function. Endoprosthetic replacement facilitates rehabilitation of older patients and a quicker healing process.

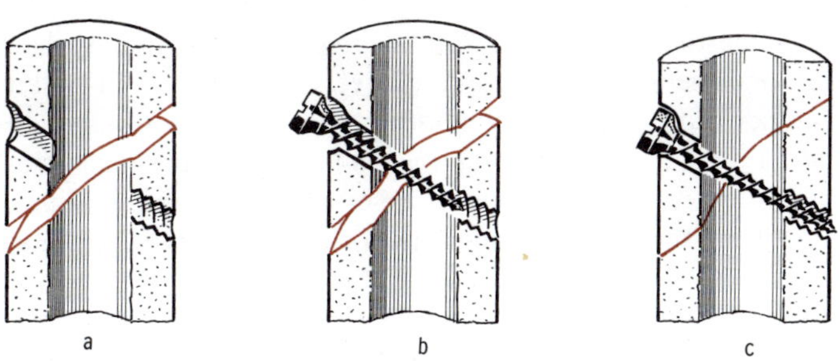

Figure 1–9 Screw fixation of fragments in long bones: a. drilling the gliding hole (left, d = 4.5 mm) and the threaded hole (right, d = 3.2 mm), b. inserting the screw, c. adaptation and compression by a cortical screw

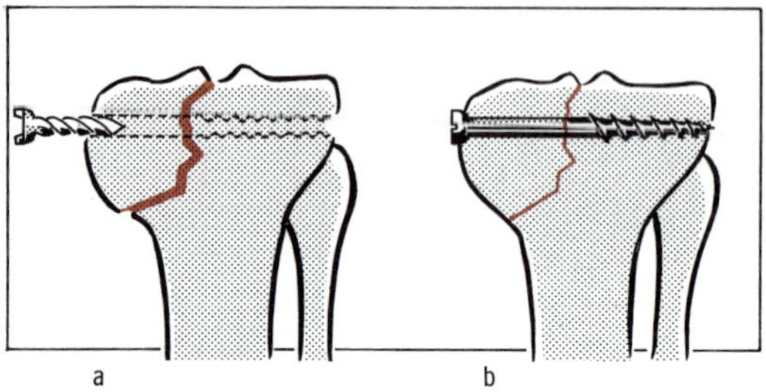

Figure 1–10 Osteosynthesis with a screw through spongy bone: a. drilling the screw hole, b. compression of the fragments with a lag screw

Osteotomy is a complete cut through a bone for therapeutic reasons; the healing process is analogous to that for a fracture. The bone is osteotomized close to the joint to utilize the favorable ossification tendency in the area of the spongy bone. The osteotomy is done with a chisel or oscillating saw after the dissection of the bone segment. One clinical indication for an osteotomy is the correction of a fracture healed in malalignment. Usually, the corrective osteotomy is followed by an osteosynthesis.

The principles of fracture repair in children are relatively simple. Alignment along the axis is desired. However, the younger the child is and the closer the fracture is to the joint, the greater the deviations that can be tolerated, since continuing growth will gradually adjust the axial deviation. A reposition in the middle third of the shaft must be accurate; however, an exact alignment of the length is not necessary, since a hyperemia of the fractured bone can cause an overgrowth of up to 4 cm. In children, the conditions for a fracture treatment without fixation are favorable, except in joint fractures, which require surgical fixation. Rotary malpositions will, however, persist and must be surgically corrected.

In children, the bony consolidation of a fracture is completed more rapidly than in adults. Immobilization is necessary only for a relatively short period, though it might still have to include more than the neighboring joints, in view of the high activity level of children. Pain is a good indicator and a limiting factor in gauging the intensity of exercise therapy for children. Over a long period of time, continuing improvements in function and use of the injured extremity in play and daily activity can be expected.

The physical therapy is always dependent on the type and the locale of the fracture. The advantages and disadvantages of the treatment must be considered with regard to the individual circumstances of the patient.

Physical therapy has to be adjusted to the type of fracture management. In fractures stabilized by osteosynthesis, active exercises are started as early as a few days after the operation; this stimulates the circulation of blood and lymph and prevents joint contractures and muscle atrophy. Physical therapy must take into account not only the local fracture site but also the general condition of the patient. Patients who are bedridden for a long period of time, especially older patients, require regular respiratory therapy, circulatory training, and prevention of decubitus. Range of motion exercises are applied to the injured extremity, and the joints that are not immobilized are moved actively to avoid contractures. The intact extremities are actively moved to support circulation. As we have noted, after an extended period of immobilization, symptoms of disturbed circulation, such as paleness or reddish-blue discoloration of the skin, may appear, together with capsular swelling and congestion of the joints (resulting in loss of joint contours), decreased range of motion, and loss of muscle substance. The reduced range of motion (especially in children) after a short period of immobilization is caused primarily by muscular problems; later on, the limitation in motion is caused mainly by problems with ligaments and joint capsules (see Figure 1–11).

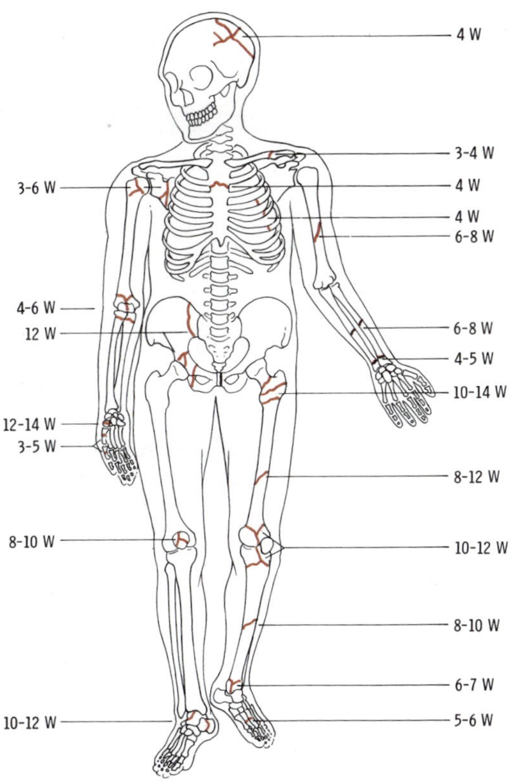

Figure 1–11 Scheme of the average healing period of fractures (listed in weeks)

Source: Schlosser and Kuner, 1980

Metal Removal

In most cases metallic devices—such as plates, screws, and cerclage wires—are subsequently removed, and external fixation devices must be removed. Internal or external metal devices are removed after the bony consolidation of a fracture or stability is established after a subluxation or arthrodesis.

The need for removal is obvious in the case of an external fixation that is protruding laterally or bilaterally from the soft tissue and causing discomfort. Also the continued percutaneous penetration increases the risk of invasion by pathogenic agents. As soon as there is adequate indication of bony consolidation, the external fixation devices are removed. Depending on the localized circumstances, this is followed by a cast or physical therapy.

Generally, internal fixation metal devices are removed after approximately one year, provided there is a sufficient bony healing. Radiographically, the cortical bone must be continuous and the medullary canal must show signs of regeneration.

In a fixation with two plates, it might be better to remove one plate first and the second one later. A plate osteosynthesis causes a certain degree of osteoporosis after the bone remodeling, since the plate takes over the load transmission and the bone starts to degenerate because of the inactivity. After the metal removal, the bone has to be loaded carefully to avoid another fracture, which is most likely to occur at the screw holes. Special care or treatment after removal of a Kuntscher nail is usually not necessary. Crutches have to be used for 4 to 6 weeks following the fracture, and physical therapy should be started early.

SPLINTS

The position of an injured extremity is an important consideration, both after trauma (for transportation, see the section on first aid) and after therapeutic manipulations.

Positioning with a splint serves two purposes: immobilization and elevation. For both purposes, the position should:

- reduce pain
- provide muscle relaxation
- improve venous drainage (to prevent edema and thrombus formation)
- limit hemorrhage and hematoma

In nonsurgical treatment a splint is applied until there is a resolution of the soft-tissue symptoms (hematoma, edema) and it is replaced by a fixation device (cast, bandage) or until mobilizing therapy begins. After surgical treatment a splint is indicated until the wound is healed, usually 10 to 14 days. Afterward, or even during the splinting period, progressive exercise therapy is recommended. In case of a persistent tendency for swelling or a risk of thrombus formation, the splint must stay in place during the physical therapy. Usually, the more distally the trauma has occurred, the higher the proximal part of the extremity must be elevated.

Besides providing a device for fixation and immobilization, splints are used for temporary positioning, particularly in the postoperative phase. The following types of splints are most commonly used:

1. Lower extremity

    - Foam splint. The extremity is bedded in an anatomically preshaped foam splint; the knee is slightly flexed (10 to 15 degrees), and the ankle is kept in a neutral position (0 degrees). This type of splint is used postoperatively, for example, in fractures close to the hip, after knee surgery, or in tibial fractures (see Figure 1–12).
    - Volkmann splint. The indication for this splint is analogous to that for a foam splint. The splint consists of a metal frame surrounding a wrapped sheet metal tray.
    - Elevated foam splint. An elevated foam splint is sometimes indicated, especially at night (30 to 45 degree flexion). This type of splint is used in femoral or tibial fractures.
    - Braun and Kirschner splint. Braun splints are made of metal frames wrapped with bandages and are flexed at the knee joint in a fixed angle.

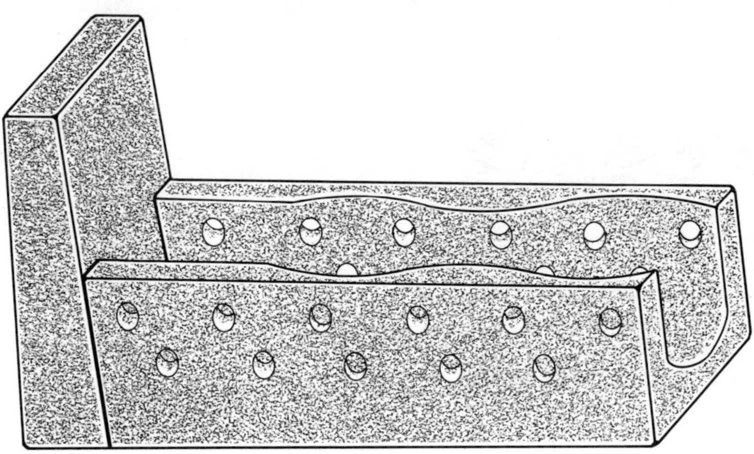

Figure 1–12 Dense foam rubber splint

Kirschner splints hold part of the extremity and have an adjustable knee angle. Modifications of these splints are also available. In these splints the foot is glued into a stocking and hooked to a frame (see Figure 1–13).

- Functional splint. Functional splints are effective in postoperative treatment as well as in early, functional active exercises following tibial head fractures with minimal dislocation. The splints are fixed to the foot end of the bed or to a bar and can be operated by pulleys (Frankfurt model) or by the patient's own muscle tension (Ulm model). They allow a passive or active knee flexion by the patient in varying degrees.

- Passive motorized movement splint. This exercise splint is electronically driven by a small computer giving a prescribed amount of passive movement at preplanned intervals throughout the day.

2. Upper extremity

- Thoracic abduction splint. With the shoulder positioned in abduction, this arm splint is tightly fixed to a ring around the chest. It is adjustable for elbow flexion. The arm is held in such a way that the inferior capsule of the

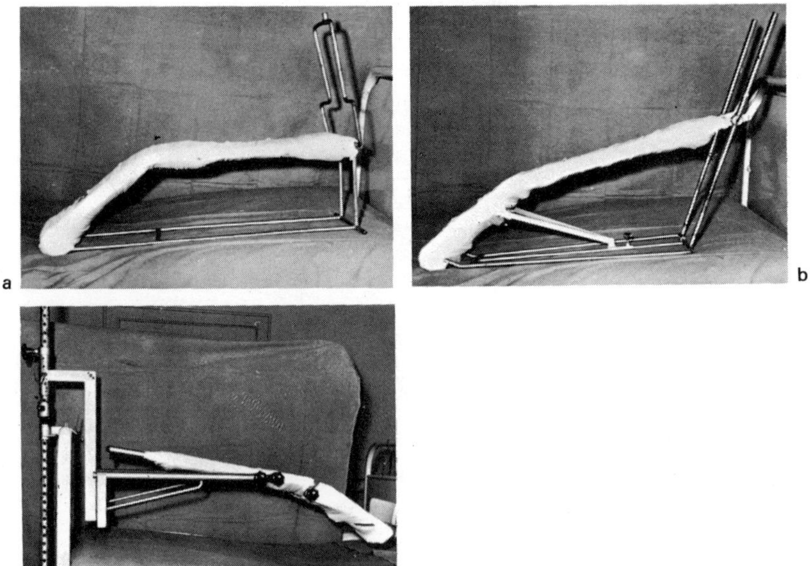

Figure 1–13 Positioning splints for the leg by a. Braun, b. Kirschner, c. Krapp

Source: Schlosser and Kuner, 1980

shoulder joint is unfolded and adhesions are avoided. This prevents the development of a permanently stiff shoulder.

- ''Wrist'' ortheses. These are plastic splints that are used to support the dorsal or volar carpals. They can be applied with various modifications, depending on the type of trauma and the functional loss.
- Finger joint extension splint. In using this splint, the ruptured extensor tendons of the finger joints will usually heal best with the finger in hyperextension for about 6 weeks (see Figure 1–14).

3. Upper or lower extremities.

In an emergency, a wire mesh splint is used for the upper or lower extremity. This splint consists of wire mesh and is held in place on the extremity with bandages. It is manufactured in several sizes, and the length can be adjusted to individual needs.

HEAD AND INTERNAL INJURIES: GENERAL CONSIDERATIONS

The severity of a head injury can vary considerably, depending on the site and the force of the impact. Soft-tissue lesions and hematomas require medical care. Professional assistance is necessary if important structures, such as an ear or eye, are involved. Extensive wounds across the skull must be treated for multiple hemorrhages to avoid an inflammation, and they must also be checked for underlying fractures.

A fractured skull can cause hemorrhage into the brain that may result in a life-threatening situation. Hematomas around one or both eyes may indicate a fractured skull with coagulated blood accumulating within the orbits. Fluid draining from the nostrils is caused by lacerations of the dura. However, because even a

Figure 1–14 Finger joint extension splint

relatively harmless facial injury can create the same symptoms, radiographs are necessary to determine possible fractures.

A concussion of the brain is a frequent injury. It is classified as a closed head trauma, as distinguished from an open skull fracture or injury associated with a brain injury. Concussions may be characterized by vomiting, abnormal pulse, and loss of memory (retrograde amnesia). A concussion is considered to be a mild brain injury with only temporary disturbances for a few days in circulation, mental activity, and the autonomic nervous system. In a brain injury resulting in a contusion, the symptoms continue for several weeks. In a severe brain injury, one can expect to see permanent mental, motor, sensory, and sometimes autonomic nervous problems (see section on special treatment).

Internal injuries caused by an impact are beyond the scope of this discussion. However, the life-threatening effect of hemorrhages from ruptured vessels and lacerated organs must be noted. Usually, the relevant symptoms become obvious only after several hours, or even days. Thus, careful observation is mandatory after an accident. Such observation is also important in the case of minor bruises in the abdominal area that happen suddenly and without any protective muscle tension.

Chapter 2

General Considerations of Physical Therapy

B. Morgenroth and U. Bergmann

Trauma care with its variety of injuries and therapeutic techniques offers a wide and fascinating field of endeavor to the physical therapist. Working with trauma patients, like any other professional interaction, requires good communication and cooperation between the physical therapist and the physicians, nurses, occupational therapists, and all others who are involved with the patient's care.

CLINICAL EVALUATION AND TREATMENT

In trauma care, the therapeutic goal is optimal rehabilitation, that is, the best possible restitution of strength and function with the minimum of pain for the patient, thereby enabling the patient to resume as near normal a level of activities of daily living as possible. The physical therapist applies the relevant expertise to help achieve this goal without unnecessary complications and delays.

A treatment program based on a careful and correct evaluation is mandatory for successful physical therapy. The more accurately the diagnosis describes the patient's problems, the more specific the physical therapist can be in preparing individual therapy plans. After the initial diagnosis, intermediate check-ups and therapeutic modifications are necessary as the situation and requirements change during the period of rehabilitation.

In the following sections, detailed information regarding therapeutic procedures is presented. Thereafter, the specific diagnoses and therapies are discussed for each type of trauma and its sequelae. In this discussion, though it is not specifically mentioned in each case, all of the indicated diagnoses must begin with the gathering of basic information and personal data.

The following evaluation outline consists of several consecutive phases, each of which involves specific observation and assessment procedures. In actual practice, rarely does a case involve only one component of the outlined evaluation items; usually, several methods are combined, and the original item is modified (by abandoning or adding items) during the course of the evaluation. At its conclusion, each evaluation must be documented in a detailed final report.

PHYSICAL THERAPY EVALUATION

The following items in the clinical evaluation for physical therapy may be expanded as necessary:

1. Personal data
    - name
    - age
    - occupation
    - date of accident
    - diagnosis
2. Basic information
    - medical history
    - surgery (type, immobilization, date)
    - radiograph findings
    - precautions
3. Observation
    - position
    - scars
    - trophic disturbances: edema, skin discoloration
    - joint deviation, intra-articular position of the bones
    - mobility, posture, gait, etc.
4. Palpation
    - sensation
    - temperature, sweating
    - muscle tone and joint displacement
    - muscle atrophy, edema
    - pulse

5. Measurements

- circumference, comparison of sides, comparison of first and following measurements
- leg length in supine and standing positions, taken with tape measure
- joint range of motion
- vital capacity
- blood pressure

6. Motor control

- muscle strength, endurance, and coordination
- disturbances in movement patterns and gait
- signs of altered reflexes, spasticity, etc.
- equilibrium reactions

7. Information from the patient

- degree and consistency of pain
- effect of the disability in personal life and occupation, etc.

8. Supplementary observations

- behavior: cooperation, comprehension, lability
- social behavior

TREATMENT PROGRAM

The following physical therapy may be employed. The sequence of application need not necessarily follow the sequence of phases in the clinical evaluation.

1. Prevention of pressure sores

- positioning
- passive and active muscle activity
- exercise in various positions

2. Prevention of pneumonia

   - positioning
   - breathing exercises

3. Prevention of thrombosis/improvement of circulation

   - venous compression with elastic bandages and/or stockings
   - elevation of extremities
   - range of motion
   - isometric and active exercise

4. Prevention of edema

   - venous compression with elastic bandages and/or stockings
   - elevation of extremities
   - massage (effleurage, intermittent kneading)
   - isometric and active exercise
   - cryotherapy

5. Circulatory support

   - changing position
   - muscle and connective tissue massage
   - hydrotherapy (whirlpool, contrast bath)
   - cryotherapy
   - electrotherapy (galvanic high voltage and inferential currents)
   - passive and active exercises

6. Mobilization

   - prevention of contractures by
     —positioning
     —passive and active exercises
   - elimination of contracture by
     —manual therapy
     —relaxation technique
     —stretching
     —ice packs (if not tolerated, possibly heat)

—connective tissue massage
—hydrotherapy
—sustained traction

7. Muscle Re-education

   - electrotherapy
   - proprioreceptive neuromuscular facilitation (PNF)
   - biofeedback

8. Strengthening

   - electrotherapy (functional electrical stimulation—FES, inferential, etc.)
   - active resistive exercises with and without equipment (PNF)
   - pool program
   - posture training

9. Training in function and movement

   - upper extremity: exercises with and without equipment, activities of daily living
   - lower extremity: gait training

PROPHYLACTIC TREATMENT AND PHYSICAL THERAPY

Prophylactic Preoperative Treatment

In the very few trauma cases in which the patient is not taken directly from the accident site to the operating room but is hospitalized first and then prepared for surgery, the physical therapist establishes the following program:

- pneumonia prophylaxis by: breathing exercises, postural drainage, positioning
- thrombosis prophylaxis by venous compression, passive and active exercises
- pressure area prophylaxis by positioning and, to the extent possible, by active movement and position change
- circulatory stimulation by active exercise
- mobilization and strengthening of the intact parts of the body, assuming immobilization of the traumatized area is secured

Prophylactic Treatment for Bedridden Patients

Prophylactic treatment of surgical or nonsurgical patients can prevent unnecessary complications, such as symptoms resulting from inactivity in the respiratory and circulatory systems, contractures, and muscle atrophy. During the initial period, the patient is instructed and assisted by the therapist. Eventually, the patient personally carries out the treatment as much as possible.

The following treatment should be implemented:

- intensification of respiratory function by breathing exercises
- circulatory stimulation by active exercises
- thrombosis prophylaxis by applying elastic wraps and active exercises
- pressure area prophylaxis by positioning, isometric and isotonic exercises, and, to the extent possible, position change
- mobilization of all joints that are not immobilized, for example, toe, knee, and hip exercises with a casted lower leg
- prevention of loss of muscle strength in all movable muscles by PNF, resistive equipment: theraband, dumbbells, pulleys

It is important to maintain regular intensive exercises. However, the exercises should not cause pain in the injured area or interfere with the healing process. Nor should they interfere with the correct bone position or result in unfavorable transmission of tensile forces or premature weightbearing.

Prophylactic Treatment of Patients with Injured Arms

Unnecessary secondary complications after injuries of the upper extremity can occur. Most commonly, they occur during or after the cast immobilization of one or more joints. These complications are characterized by decreased blood circulation, function, and strength in the adjacent parts of the extremities and joints.

To avoid these complications, the patients are given instruction in the following exercise program:

- elevation of the injured arm on a wedge pillow
- exercise of nonimmobilized joints such as shoulder or fingers
- temporary elevation of the injured arm while standing up and moving around; leaving the arm on the lap, armrest, or table while in sitting position
- movement and exercises in the cast, to the extent possible
- exercises for the fingers with a foam ball or other small items
- activation of the elbow joint if in a forearm cast

- movement of the shoulder in all directions if not immobilized, in prone position and while sitting and standing up; at first, supporting the weight of the arm and later unsupported movement

Through these exercises, the patient should become convinced that the arm can be activated without aggravating the injury. It must be emphasized to the patient how important these exercises are and how regular exercise furthers the healing process.

THERAPEUTIC CONCEPTS INVOLVING FRACTURES

Exercise and Weightbearing Stability

In the case of an unstable injury, therapy should be begun in close consultation with the physician. One starts, very carefully, with isometric exercises. If the patient is in an unstable condition, the physician might recommend specific precautions, for example, to support the weight of the extremity and avoid certain directions during movement. If there are no such restrictions, the patient may be able to do active exercises in each plane of movement.

Exercises that allow resistance below the fracture or that involve stretching, traction, or pressure to the fracture site must be avoided until the fracture is consolidated, regardless of whether the fracture had been treated surgically or nonsurgically. As soon as the fracture appears on radiograms to be consolidated, isometric and dynamic resistive exercises using equipment can be started.

It is clear from the above discussion that exercises are almost entirely active. Seldom is it necessary to do passive exercises. The currently dominant method of fracture fixation, ORIF (open reduction, internal fixation), creates osteosyntheses that are stable enough for active exercises, thereby avoiding severe contractures. At the same time, over the years, more precise methods of exercise have been developed, such as PNF techniques, which can be modified for specific situations or combined favorably with cryotherapy.

Apart from the greater efficiency of active exercise, the risk of loosening (fixation material) is greatly reduced. However, in the case of old and persistent contractures that have not responded to extensive active exercises, and where the fracture site is obviously stable enough, passive mobilization therapy might be implemented.

Weightbearing for Gait Training

Weightbearing in a Therapeutic Pool

The use of a therapeutic pool is especially advantageous in the case of patients who have to be confined to their beds (without weightbearing) for long periods of time. At first, weightbearing is minimal, with the water level up to the shoulders and neck. Gradually, the water level is lowered, or the floor is elevated. The physician provides information on how rapidly weightbearing should be increased and at what point the patient could start weightbearing outside the water.

Nonweightbearing or the Use of "Toe-Touch" Weight

Since it is frequently recommended in the literature that a "professionally stabilized osteosynthesis" be subjected to early partial weightbearing (Muller et al., 1969), the following method is suggested: usually, the patient is standing and supported with crutches and nonweightbearing or while slightly touching the ground with the toe. The weightbearing force at the ground should be in the range of 5–10 kg. The physical therapist must make sure that the patient who is being allowed to touch the floor understands the prescribed weightbearing and follows it precisely. If there are any doubts, walking nonweightbearing with crutches should be maintained.

Estimating the Weightbearing

Exact physical data about the dynamic forces that can be safely applied to the human skeleton cannot, of course, be calculated for each individual patient. Therefore, the static load on the extremity, an easily accessible datum, is selected as the basic information for determining the amount of weightbearing. In the following paragraphs, two methods that are well established in daily clinical use are described.

Method 1. The patient stands still with the intact leg on a lift that is the same height as the scales (see Figure 2–1a). The crutches are adjusted accordingly on the floor, or they are also put on lifts. The injured leg touches the scales. The therapist sets the scales to the desired weight, and the patient puts weight on the scales gradually until the selected number of kilograms or pounds is reached. The patient is asked to maintain the weight during the weightbearing training. If successful, the patient should then try to push the weight again with eyes closed. Afterward, the patient applies the newly acquired sense of weightbearing in a three-point gait on crutches.

Method 2. A less troublesome and more quickly executed technique is to do the weightbearing training in an easily arranged contrived setting (see Figure 2–1b). Commercial scales (calibrated) are set into a wooden frame 20 to 30 cm wide and the same height as the scales. The patient is able to step comfortably non-weightbearing onto the frame, and then put enough pressure on the scales to achieve the desired degree of weightbearing. To ensure the correct amount of partial weightbearing, the patient should repeat the procedure several times during

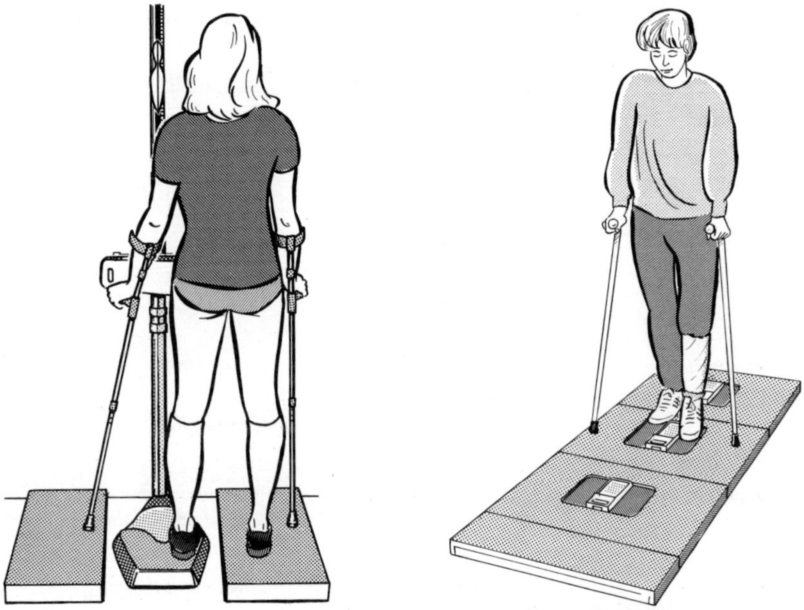

Figure 2–1a and **2–1b** Methods of estimating the load

the day and, if necessary, correct any deviation. The degree of weightbearing appropriate for particular types of equipment and gait patterns are listed in Table 2–1.

Additional Aids

In fractures of the lower extremity, the patient needs an arch support or other orthotic supports if:

- there are already foot deformities
- there is a risk of secondary ankle or foot problems due to the location and type of fracture
- there is secondary demineralization of the metatarsals due to the long immobilization period

Elastic stockings may be necessary after injuries of the lower extremities. Function may be improved with better circulation and soft-tissue support.

Table 2–1 Type of Weightbearing, Auxiliary Means, and Gait Pattern

Type of Weightbearing	*Auxiliary Means*	*Type of Gait*
Nonweightbearing	parallel bars, walker crutches (Figure 2–2)	swing through
toe touch weightbearing	parallel bars, walker, crutches	3-point gait with toe contact: 5–10 kg
partial weightbearing	crutches	3-point gait
full weightbearing	two crutches or two canes (Figure 2–3)	4-point gait 2-point gait
full weightbearing	one cane (on the normal side)	2-point gait
full weight	none	normal gait
For stairs and curbs		
Non, partial, or full weightbearing	rail, crutches, or canes	upstairs: normal leg first downstairs: injured leg first

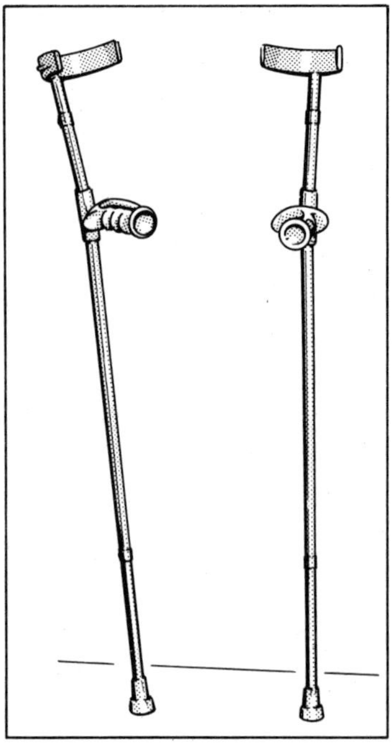

Figure 2–2 Light metal crutches with anatomical handles

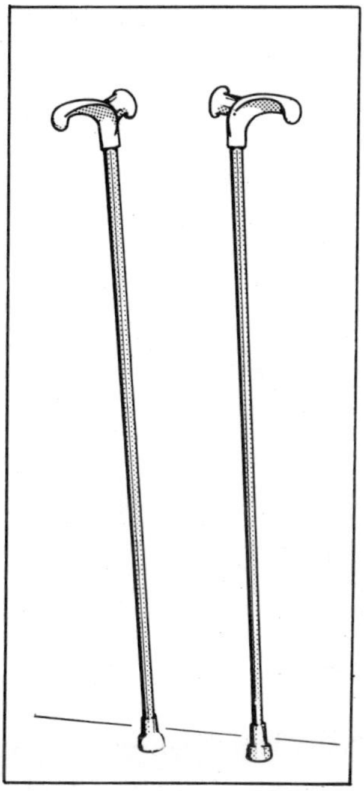

Figure 2–3 Light metal canes with anatomical handles

PRINCIPLES OF PHYSICAL THERAPY IN MULTIPLE TRAUMA PATIENTS

In traumatology, the physical therapist is confronted increasingly with multiple trauma patients. Despite the lack of standardized rules for physical therapy because of the variety of injuries and their combinations, some basic tenets for physical therapy with multiple trauma patients can be formulated.

First, good cooperative relationships within the therapeutic team (physician, psychologist, social worker, nursing staff, physical therapist, occupational therapist, and so on) are important.

Second, for such injuries, not only is a fundamental professional knowledge necessary, but also a readiness and ability to provide psychological support.

Third, the prevention of pressure areas, pneumonia, and thrombosis and circulatory stimulation are initially the most important concerns. As the ability of the patient to endure stress and exercise increases, attention is shifted to mobilization, strengthening, and daily exercises.

Fourth, because the patient is usually confined to bed for a relatively long period of time and thus suffers from boredom and limited activity, the physical therapist should offer a variety of exercise procedures and equipment to stimulate and maintain the patient's motivation in each phase of the treatment. This can be done through the use of interesting modifications during the exercises and by providing daily rewards for incrementally successful results. Excessive demands that lead to frustration should be avoided.

Finally, physical and occupational therapists should be ready and able to improvise. For example:

- To increase independence in activities of daily living, therapists should determine which aids would be most useful: adaptive utensils for eating, braces, book and newspaper holders, page turners, reachers, prismatic glasses, and so on.
- Therapists should provide therapy at the bedside.
- If the patient's hand or forearm is too weak for self-support, additional support should be provided under the arms, but the patient should be observed for any sign of damage to the nerves from the pressure applied.
- With patients who are limited to partial weightbearing bilaterally, walking exercises should be done in a pool with a high water level.

In the treatment of patients with multiple trauma, it is most important not only to see the combination of injuries as a clinical case but also to see the patient as a human being and to consider that person's needs in each phase of rehabilitation.

NOTES ABOUT SELECTED TREATMENTS

Prophylaxis and Treatment of Pressure Sores

The risk of pressure sores is small in patients who have been treated immediately with a stabilizing osteosynthesis and are able to get out of bed within a short period of time. However, the physical therapist should always ensure that the usual postoperative bandages and splints do not cause pressure, that the patient is resting comfortably, and that critical areas, like the heel or sacrum, can be protected with a cushion or sheepskin or relieved by simple exercises, by isometric contractions of adjacent muscles, or, if possible, by dynamic movement or changing positions. When the patient is confined to bed for a long period of time, these critical bony areas must be watched with particular care by the physical therapist.

Older multiple trauma patients and patients with insufficient circulation or with disturbances of the sensory or autonomic nervous system are especially at risk. If a pressure sore develops in spite of all precautions, the following therapeutic procedures should be used, together with continued prophylactic care:

- application of ice packs or hot air to the peripheral area of the sore
- connective tissue massage in the area of the pressure sore

Ice Therapy

Treatment with ice packs is an important part of trauma care. It is indicated for painful muscles and joints, inflammations, and facilitation in exercise techniques. It is also used as a means to stimulate muscles with paresis following fractures with neurologic involvement.

The symptoms will determine the way in which the ice is applied: stroking, ice massage, ice towel, ice pack, ice bath. The selected method should never be used alone; it should always be combined with exercises. In no case should the application of ice interfere with the healing of a wound by increasing localized moisture.

Hydrotherapy

A therapeutic pool is a valuable asset in rehabilitation (see Figures 2–4a and 2–4b). The pool can be used to support the circulatory function, mobilization, strengthening, and gait training. In addition, it has a distinctly positive analgesic and psychological effect in nearly all patients.

Ideally, the pool should have an adjustable floor or a sloping height floor to enable the physical therapist to choose the desired water level of an individual treatment. In order to expand the range of exercises or assist weak muscles, the exercises should be designed in a way that either utilizes the water's buoyancy (motions are directed toward the surface) or neutralizes it (motions are directed horizontal to the surface). Patients can be advised to work against the buoyancy in order to strengthen their muscles intensively (motions are directed toward the pool's floor). In this case, the fracture should be stable enough to withstand the resistance and possible shearing impact of the water. Special equipment can be used to provide additional support or to facilitate resistance to the buoyancy, in combination with exercises. A pool with a deep floor and high water level is very helpful in the first partial weightbearing exercise. A patient who has been trained in such a pool has less trouble doing the exercises outside the water.

The therapeutic pool also offers opportunities for individual treatment. This sometimes requires that the therapist go into the water with the patient. Alternatively, patients with similar diagnoses and comparable abilities can exercise together in the pool. In such cases, the therapist could observe from the poolside, announcing and correcting the exercises as necessary.

Figures 2–4a and **2–4b** Therapeutic pool

Electrotherapy

Electrotherapy that is used in trauma care primarily involves the use of intermittent or galvanic currents. Electrical currents are used to support the resorption of hematomas and effusions. Galvanic currents are used to avoid muscle atrophy by stimulating contractions in a neurologically affected muscle after a fracture. Intermittent currents help to strengthen weakened muscle.

Electric currents should be judiciously applied, since they can cause chemical, thermal, or other damage to the osteosynthetic material in the areas of the electrodes.

Exercises in Front of a Mirror

In nearly all exercises that patients do while sitting, standing, or moving around (regardless of the type of injury), it is helpful to work with the patients in front of a mirror. In this way they are able to control themselves visually and to train and improve their conscious awareness and the direction of their motions.

Independent and Home Exercises

All patients should be motivated to do their own exercises outside of scheduled therapy. The physical therapist should always have an instructive, though not overly extensive, home program ready for outpatients.

Additional Therapeutic Possibilities

Various therapeutic methods, such as swimming, saunas, occupational therapy, and recreational sports, are mentioned throughout the remainder of this text.

Part II

Specific Trauma and Treatment

Chapter 3

Head Trauma

W. Heipertz, L. Zichner, B. Morgenroth, and U. Bergmann

SCALP INJURIES

Mechanisms of Injury

Injuries of the skin covering the skull are caused by a direct force. This tissue is well vascularized; it has a good healing tendency and becomes infected only in exceptional cases.

Scalp injuries are treated following the general principles discussed in Part I. If excisions are necessary, they must be done sparingly, and the embedded structures must be spared if possible. Scalp injuries may be associated with skull fissures, which do not show up radiographically. Eyeballs and ear cartilage must be checked. Cosmetic suture techniques are required in the area of the eyebrows, eyelids, ears, and lips. Adrenalin should not be added to the local anesthetic.

Local wound infections of the scalp can be complicated by an inflammation progressing to the skull (ostitis). From the facial bone, the infection can invade the sinus cavernosus and can cause the formation of thrombi.

FACIAL BONE FRACTURES

Mechanisms of Injury

Fractures of the facial bones are caused by a direct force. They can affect the sinuses (maxillary, frontal, and orbital). These fractures affect primarily the nasal bone, the zygomatic arch, and the zygomatic bone.

Types

Fractures in the maxillary area are classified according to Lefort in one of three groups, depending on their extent. Mandibular fractures are variable.

Clinical Concerns

Fractures of the facial bones are accompanied by deformation (deviation of the nose and ocular axis), pain, swelling, functional limitations, and frequently hematoma. There is often a dental malocclusion, which requires treatment in consultation with specialists.

SKULL FRACTURES

Bone lesions of the skull and brain trauma pose special anatomical problems. In cases of hemorrhage or edema, the brain and increased cerebral fluids enclosed by the bony skull will give way only to a limited extent. The intracranial pressure increases inside the skull and is manifested by a dilated fixed pupil (pressure on the oculomotornerve). Finally resulting in severe damage to the brain, this process is called decerebration. A paralyzed respiratory system and circulatory collapse can be fatal. Thus, careful observation and early intervention are mandatory after skull-brain trauma.

Fractures of the skull can be categorized as fractures of the vault and/or of the base. Vault fractures have significant additional complications, particularly brain damage or vascular lesions, which cause epidural or subdural hematomas.

Special Fractures

Depression fractures are characterized by an inward displacement of fragments into the skull. Immediate operative elevation or removal of the fragments is mandatory. If a penetrating fracture opens up the intracranial space, fluids can drain and brain substance can protrude. The risk of infection in this case dictates immediate surgery, with repair of the meninges. In any case, a cranial vault fracture makes hospitalization necessary for observation and early detection of any complications.

Fractures of the cranial base are difficult to recognize. More often than not, they are verifiable by indirect symptoms rather than radiographs. Fluids drain out of the ear, nose, or mouth. Air accumulation and a positive radiograph are definite indicators. In contrast, hemorrhages from the mouth, nose, or ear and uni- or bilateral periocular hematomas can occur without a base fracture. Cranial base fractures require hospitalization for observation and early recognition of further complications. Infection is possible in the presence of a fluid fistula. An early complication is meningitis. Later complications are brain abscess and sinus thrombosis. Therefore, antibiotics are given prophylactically.

BRAIN TRAUMA

Mechanisms of Injury

A skull and brain trauma can occur in addition to the injuries described above but also on its own. Brain trauma without accompanying soft-tissue or bone lesions is called a closed head trauma.

Types

A cerebral concussion, contusion, and compression are each characterized by their individual symptoms and durations. Recently, each has been classified in terms of three degrees of damage, depending on the duration and elimination of the symptoms. A correct estimation of the degree of the lesion is therefore possible only retrospectively.

A cerebral concussion is considered a first-degree lesion. It involves functional disturbances that subside after a few days. There are no structural changes, and no permanent sequelae ensue. The initial unconsciousness lasts only a few minutes and may be followed by vomiting, memory loss, and headaches. After a concussion, the patient should be observed for two days and confined to bed rest. If no complications occur, regular activity can be resumed.

The symptoms of a cerebral contusion (second-degree lesion) are analogous to those of a concussion, except that they last longer. The symptoms subside within three weeks. The brain substance is damaged locally at the site of the impact (coup) or on the contralateral side (contre coup).

In a severe cerebral compression (third-degree lesion), the symptoms often persist much longer. The destruction of brain substance results in permanent defects. Massive subarachnoidal hemorrhages cause acute subdural hematoma and cerebral compression. Craniotomy is necessary to stop the hemorrhage and evacuate the hematoma.

Medical Management

Intubation, to keep the airway patent, and oxygen for circulatory support are the most important first-aid procedures. Nonsurgical observation and diuretic therapy are essential. Persistent unconsciousness requires alimentation parenterally or by stomach tube. Episodes of restlessness are relieved by tranquilizers. Elevated temperatures are treated with ice packs and antipyretics.

Complications

Decreased pulse and temperature elevation may indicate increased cerebral pressure. Additional symptoms are intensifying headaches, restlessness, nausea, disorientation, neurologic deficits, repeated unconsciousness, and dilated pupils. The deficits are dependent on the continuation and magnitude of an epi or subdural hematoma or cerebral edema.

Later Complications

Spasticity and paralysis may remain after a cerebral contusion or compression. Pyramidal symptoms, cranial nerve paralysis, different pupil sizes, and fever are observed clinically.

Differential Diagnosis

The situation after a closed head trauma resembles that of alcohol abuse. In most cases of a suspected head trauma, hospitalization is necessary.

Physical Therapy

The following therapeutic program is applicable in the case of a cerebral concussion and skull fractures without significant neurologic deficits:

Symptoms

- headaches
- nausea
- vomiting, especially after abrupt movements, position changes, and attempts to rise

Therapeutic Aspects

- pneumonia and thrombosis prophylaxis: (1) respiratory therapy, (2) isometic and dynamic exercises
- circulatory training: (1) active movements, endurance training, (2) changing into different positions
- mobilization and training of basic skills: (1) mobility exercises, (2) active exercises work in different patterns, (3) exercising in various daily activities

Course of Treatment

During the immobilization phase, which usually lasts only a few days, pneumonia and thrombosis prophylaxis and circulatory training are the main therapies. Any increased respiratory frequency must be reduced, and ventilation must be improved. The coughing up of secretions or mucus is encouraged. The extremities are moved actively, from distal to proximal joints, ultimately vigorously and accelerating the tempo (endurance training).

As soon as a change of position is allowed, the physical therapist begins to train the patient to turn from the back to the side and onto the abdomen and to sit and stand up. In each new position, stabilizing exercises are done. The positional changes are initially done slowly and then more and more quickly. Thus, at the time the patient begins to stand up and walk around, the circulation becomes adjusted to the changing positions. If the frequency of the activity is insufficient or its rhythm becomes irregular, particularly in sitting and standing positions, the therapist should observe the patient's pulse and blood pressure and adjust the activity accordingly.

Training in daily activities, for example, looking back over the shoulder, looking and reaching up, bending down to tie shoelaces or to pick up things from the floor, is the next phase in the therapeutic program. Usually, the patient will feel self-confident and secure again within a few days and can be discharged from the program.

Sample Exercises

The following exercises may be used for changing positions:

1. Turning from a supine position onto the side. The patient lies in supine position with flexed legs and turns (with head raised and feet dorsiflexed) against the therapist's controlled resistance on shoulder and pelvis onto the side and then again into supine position. This is done at first slowly, then more quickly, with changing tempo, depending on the patient's ability.
2. Turning from the side into a prone position. The patient lies on the side with extended legs and turns (with elevated head, extended neck, and feet plantar flexed) against the therapist's controlled resistance (as in Exercise 1) into a prone position and then (with elevated head, extended neck, and feet dorsiflexed), against the same controlled resistance, back again onto the side.
3. Turning from the side into a sitting position. The patient lies on the side with flexed legs, close to the edge of the bed. The patient uses the underlying arm (first elbow, then hand) and the hand of the upper arm for support and then raises from the bed, at the same time letting the legs slide over the edge of the bed. The patient then pushes and, through leverage, achieves an upright sitting position. During the entire procedure, the patient's eyes follow the changing position of the feet, then are directed forward at the therapist as soon as the sitting position is reached. The feet are set on a supporting surface or on the floor. The therapist can work with the patient sitting with or

without controlled resistance on certain body areas. The therapist may also instruct the patient to get up at first slowly, then more quickly and with changing tempo.

4. Stabilizing the sitting position. The therapist stabilizes the patient's sitting position by lateral, posterior, anterior, and rotary contacts on the trunk, head, and extremities, first on single areas (head, shoulder, knee), then in combinations (head/shoulder, head/pelvis, head/arm, knee/foot). The patient is stimulated to sit upright in any case by the feet pushing against the floor and by the therapist's hand giving pressure on the back of the head.
5. Standing up. The patient sits on the edge of the bed and the feet are placed on the floor with one foot in front of the other. The patient then shifts weight onto the leg in front and pushes up into an upright position. The other leg is then positioned beside the front leg. While standing up, the patient can gain support with hands on the bedside or with pressure from the therapist's hands and then get up. The patient also can try this without using the hands or by initially pressing against the therapist's controlled resistance on the patient's pelvis. Initially, the patient should stand up with legs astride (enlarging the base of support) for greater stability and should be able to maintain balance before starting walking exercises.

Chapter 4

Spinal Injuries

GENERAL REVIEW

The spine not only supports the body in an upright position, it also protects the spinal cord and nerves. The spine consists of segments that include the structures of bones, disks, ligaments, and muscles.

Mechanisms of Injury

Spinal injuries are caused by direct or indirect forces that exceed the resistance capacity of the structures. They usually involve a combination of injuries, affecting vertebrae, ligaments, and disks. Except in hyperextension trauma, intact disks are more resistant than bones. The soft tissue is affected in various ways. Longitudinal contusions to the back caused by falls from great heights and excessive flexion or extension can cause lesions to the spine. In such injuries, torsion mechanisms are frequently involved.

Types

Spinal injuries are classified as follows according to Lob:

1. ligament injuries
2. isolated disk injuries
3. isolated vertebral fractures
4. vertebral fractures with disk injuries
5. injury of the entire spine (fracture of the vertebral body; injury of the disk, vertebral arch, and process; ruptures of the ligaments and muscles)
6. vertebral dislocations

Soft-tissue injuries along the spine (bruises, sprains, ruptures) occur frequently but are not as problematic as vertebral dislocations or fractures. However, hemorrhages into the muscles often cause severe pain. A special form of torsion injury is the so-called whiplash injury, which occurs most frequently to the cervical spine.

Injuries of the spinal ligaments and the intervertebral disks are often difficult to evaluate and require careful analysis of the accident. The symptoms of spinal injuries are often not obvious, since they may be overshadowed by primary shock and other acute symptoms. Unilateral or complete bilateral subluxation or dislocation of the vertebrae is a clear indication of ruptured ligaments. Abrupt movements in the sagittal plane can cause bilateral or incomplete subluxations. An additional rotary force can result in unilateral subluxations.

An isolated disk rupture or fracture of the vertebral arch does not occur very often. However, both of these injuries must be regarded as serious because of their effects on the spinal cord. Fractures of the vertebral bodies are classified as (1) fractures of the posterior edge, (2) anterior depression fractures, (3) compression fractures, and (4) comminuted fractures (with anterior and posterior fragments).

Fractures of transverse and spinous processes are less severe. Isolated fractures of the transverse processes are the result of a direct or indirect force, especially in the area of the lumbar spine. With an indirect force, fracture fragments are usually torn off by uncoordinated muscular pull. Such fractures may also be caused by direct impact or a fall. Fractures of transverse processes often heal with the formation of a pseudarthrosis, which does not cause major problems.

Fractures of the spinous processes are also caused by direct or indirect impact. They usually do not require immobilization. In the area of the last cervical and first thoracic vertebrae, fractures may occur as a result of stress fatigue.

The severity of spinal trauma is usually determined by the severity of the force. In osteoporosis, even a little force can cause considerable vertebral deformation, as well as diseases, of bones, joints, and muscles, which can exacerbate the injury.

Diagnosis

The diagnosis of a spinal injury is verified by radiographic examination. The isolated fracture of a vertebral body is characterized by a deformation of the body or a density of the trabecular bone adjacent to the disk. Injuries of the ligamentous apparatus, such as subluxations or complete dislocations, are also diagnosed by radiographs, but are not always apparent. Early recognition of ligamentous injuries is rare; when they are recognized, it is usually by indirect signs (deviation of the tracheal airspace). Frequently, such injuries are discovered only when a subsequent controlled radiograph reveals a calcification in the area of the lesion, thereby indicating an earlier injury. Strains are frequent in the area of the cervical spine, whereas many of the vertebral fractures occur in the area of the last thoracic and first lumbar vertebrae.

A subluxed or dislocated vertebral injury is characterized by a bending of the spine and axial deviation. A partial or complete paralysis indicates an involvement of the spinal cord. Hypertrophic callus formation can compress the cord progressively and requires surgical intervention. Irreversible damage of the spinal cord after a dislocation or fracture of the spine is a major potential danger. Even a spinal injury without major deformation can damage the spinal cord, particularly through post-traumatic hemorrhage or edema. Hematomyelia and contusion of the spinal cord have a fairly favorable prognosis. However, a contusion or compression of the cord is likely to cause a permanent paralysis, with little chance of recovery.

Contusions (bruises) of the soft tissue usually do not result in permanent lesions, but may require physical therapy. They are caused by preexisting diseases (disk disease, wear of the vertebral facets, muscle weakness from inactivity). In case of a suspected spinal cord injury (permanent paralysis), first-aid procedures have to be designed to avoid further damage.

Medical Management

Vertebral fractures require immobilization and/or internal stabilization. Because of the prominence of forward bending, flexion injuries are the predominant result of a compression fracture. For vertebral fractures with distinct flexion injury, various treatments have been proposed. Magnus rejects a straightening alignment of the wedge-shaped vertebrae. Bohler proposes that the patient be elevated in a hanging position and immobilized in a lordosing cast. In younger patients with a downward displacement of the anterior arch by more than half, an osteotomy should be considered to elevate the vertebral body. Surgery might be indicated in subluxation fractures with neurologic deficits.

External fixation may be sufficient to stabilize the spine after the pain has subsided (see Figure 4–1). However, strict bed rest is required for a fracture with an unstable spine. After three months, a vertebral fracture is usually considered to be stable. Permanent instability of spinal segments may require a supportive brace.

Physical therapy for spinal injuries is directed at a complete restoration of function, especially the return of stability without pain. Painful attempts to increase range of motion after a sprain or torsion injury can exacerbate the injury and require temporary immobilization of the neck in a soft or hard collar.

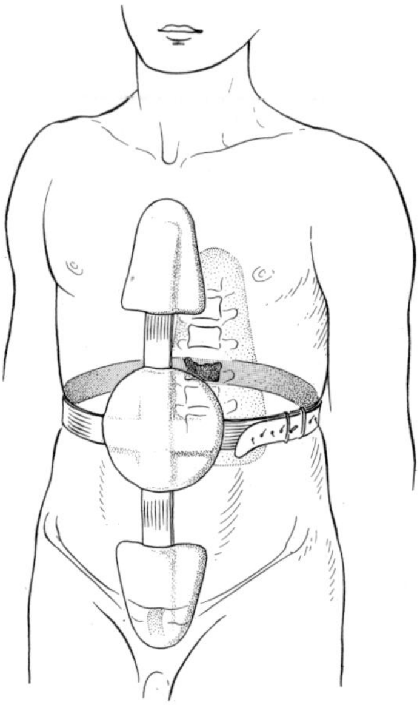

Figure 4–1 Three-point stabilization brace to stabilize vertebral fractures

Source: Schlosser and Kuner, 1980

FRACTURES OF THE CERVICAL VERTEBRAE

Injuries and fractures of the cervical spine are common; they can occur in a variety of ways because of the structure and position of this part of the spine.

Mechanisms of Injury

Fractures of the first and second vertebrae (comminution fracture of the atlas or dens) are caused by a vertical impact, such as a fall or blow on the head. Injuries of the middle and lower sections of the cervical vertebrae may also result from a vertical force, but they occur more typically from a hyperextension or hyperflexion mechanism.

Types

In a compression fracture, the vertebra appears to be deformed into a wedge shape and there may be additional subluxation of the upper vertebral joints. At the same time, the vertebral arch can be fractured and the dorsal ligaments can be ruptured. Disk lesions are characterized by a narrowing of the intervertebral space.

Clinically, fractures of the first and second cervical vertebrae cause pain in the back of the head, a stiff neck, or a weakness when attempting to hold the head up. A kyphotic flexion results from the fractures of the middle and lower cervical spine.

Medical Management

The choice of treatment depends on the extent of the instability and the neurologic symptoms; usually, the treatment is nonsurgical. Stable fractures are immobilized in a head-chest cast for three months. In cases of a dislocation and instability, Crutchfield tongs are used to align the cervical vertebrae. Surgery is uncommon (confined to repositioning and fusions in cases of increasing neurologic symptoms). The immobilization period for a vertebral fracture of the cervical spine is three months. A head-chest cast is applied if the anterior and posterior ligaments are ruptured.

A series of radiographic observations of the cervical spine injury determines the period of immobilization, depending on consolidation of the fracture by the formation of bony bridges, and so on. Fusion might be considered if the instability persists.

Physical Therapy

The following therapeutic regimen is applicable in the case of a fracture of the cervical vertebrae.

Symptoms

With the patient in a supine position and cervical stabilization achieved through use of a collar, head-chest cast, or tongs, the following symptoms are observable:

- diminished circulation in areas with constant bed contact
- muscle weakness
- possibly decreased respiratory and circulatory function
- pain at the fracture site

Therapeutic Aspects

- pressure sore prophylaxis: (1) position changes, (2) training in importance of position changes, (3) isometric and dynamic exercise
- pneumonia and thrombosis prophylaxis: (1) breathing exercises, (2) active exercises

- circulatory stimulation: (1) isometric muscle contraction, (2) active movement
- mobilization during the consolidation period: (1) maintaining the free-joint motion of all extremities by active movement, (2) lying on the left and the right sides, with active movement of the extremities from these positions
- mobilization after the consolidation is completed: active movement of the cervical spine in different directions
- strengthening through isometric and resistive exercises of the muscles of the extremities, trunk, and neck in different positions
- training in activities of daily living: (1) improving the limited range of motion by exercising with equipment, (2) exercising activities of daily living

Course of Treatment

The patient is confined to bed in the supine position. The cervical vertebral fracture is immobilized in a collar, head-chest cast, or tongs. At this stage, exercises to prevent pressure sores, pneumonia, and thrombosis are most important. Also, efforts should be made to prevent loss of joint mobility and to strengthen the muscles of the trunk and extremities. Beginning with the onset of treatment, the physical therapist may have the patient do isometric neck exercises, with the head and shoulders flat on the bed.

As soon as the side position can be achieved with the neck stabilized, the extremities are moved dynamically and the neck muscles isometrically. The head and neck must be well supported.

Beginning in about the seventh week, the patient is allowed to sit up and stand up (using elastic stockings). The patient turns and pushes to sitting against the therapist's controlled resistance (see exercises in section on skull fractures).

After the sitting position is stable, dynamic exercises can be started, with or without equipment. The patient achieves a standing position by having maintained muscle strength. The standing position is then corrected and stabilized, and walking is begun.

Beginning in about the thirteenth week, the external stabilization device can be removed. The cervical spine must now be carefully mobilized. The neck muscles can be exercised selectively in different directions or in combined exercises, as with PNF. The typical stiff neck posture is best eliminated by exercises in different positions, training in daily activities, and by walking and moving to music. Exercising in standing and sitting with the use of stall bars also can stimulate the patient to rediscover and utilize the full range of motion of the cervical spine (see Figure 12–12).

Sample Exercises

The following exercises may be used in the late phase of mobilization and strengthening:

1. PNF patterns without or with only minimal resistance and with careful rotation:

   - combined exercise for head and arm from supine position
   - "chopping" and "lifting" from the supine or sitting position (with extended legs)

2. Exercises with a ball:

   - moving the ball with both hands in different directions (away from self, around self, up and downward) with visual contact on the ball, with patient in sitting position

3. Exercises such as mimicking wall washing while maintaining visual fixation on the direction of arm movements.

WHIPLASH INJURY

Mechanism of Injury

A whiplash injury of the cervical spine is typically caused by a special kind of accident. It usually involves a driver of an automobile who experiences a direct impact from behind (see Figure 4–2). There are no external signs of the injury, but shearing forces from the rapid flexion and extension movement of the neck result

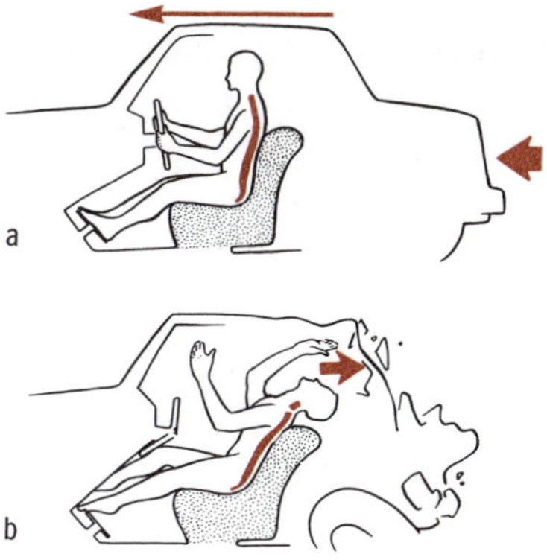

Figure 4–2 Mechanism of whiplash injury. a. after the impact from behind, a forward motion of the head; b. consecutive backward catapulting of the head in combination with a shearing force

Source: Schlosser and Kuner, 1980

in segmental lesions. Forces to the neck coming from a frontal direction are less traumatic.

Types

The fractures, sprains, and ligament ruptures described earlier can also occur in a whiplash trauma. Usually, the third to seventh cervical vertebrae are involved.

The lesions in whiplash injuries are classified according to the involvement and the nature of the follow-up problems (Erdmann):

1. Minor soft-tissue injury, such as a strain or sprain as in a contusion of the vertebral arch or small pericapsular or muscular hemorrhages. With no bone lesion, the complications subside within a few weeks.
2. More extensive soft-tissue injury, including ruptures of the ligaments and joint capsules, disruption of the disk within the segment involved, and temporary disk dislocation; with no bone lesion.
3. Extensive soft-tissue injury occurring in combination with a vertebral fracture with obvious malposition as noted on radiograms.
4. Most severe injuries match those of a subluxation fracture, including all accompanying symptoms.

Clinical Considerations

Immediately after the accident (seldom) or within the following 24 hours (frequently), persons with whiplash injury complain of pain around the back of the head and in the neck, limited range of motion of the cervical spine and nausea. Among these symptoms, the painfully limited range of motion of the cervical spine, tension and pain in the shoulder and neck muscles, and swelling of the soft tissue in this area are clinically dominant. Neurologic deficits, such as paresthesia, hypo- and hyperesthesia, and the loss of motor function, are accompanying complications of these injuries.

Diagnosis

Radiographic signs are usually not indicative, except for a frequently observed deviation of the cervical spine and a slight dislocation of the vertebrae against each other.

Medical Management

The first-degree whiplash injury might require a soft collar. The second-degree injury requires a plastic collar. In third- and fourth-degree injuries, the treatment depends on the neurologic complications and the extent of the subluxation. A cervical-thoracic orthosis is applied after the subluxation is reduced.

The physical therapy accompanying and following the injury can be done effectively by manual mobilization or mechanical traction. However, manual manipulation is contraindicated in the case of hypermobility, which requires isometric training of the supporting muscles.

Physical Therapy

The following therapeutic program is applicable in cases of whiplash trauma.

Symptoms

With the patient stabilized with a collar, the following symptoms are observable:

- pain in the area of the cervical spine
- headaches
- nausea
- severely limited range of motion of the cervical spine

Therapeutic Aspects

- circulatory support: (1) relaxation techniques, (2) massage, (3) dynamic exercises, (4) application of ice packs
- mobilization: (1) contract-relax or hold-relax techniques and application of ice packs, (2) traction, modalities, and exercises in daily activities

Course of Treatment

After a whiplash, the cervical spine is usually immobilized for about two weeks by a plastic collar. During this phase, the patient should be advised to work on circulatory support techniques.

After two to three weeks, the collar is removed. For a patient with strong muscular hypertension, the physical therapist can apply relaxation techniques. This is followed by gentle massage of the shoulder and neck, massage of the connective tissue, and active exercise in all directions of the cervical spine, using ice packs and with the patient lying in the supine position or sitting up.

Careful extension range of motion is begun and eventually strengthening of scapula, neck, and neck-arm using PNF patterns. Exercises in various positions with equipment and training in daily activities complete the program. The physical therapist should pay close attention to all exercises concerned with strengthening and with daily activities, with a view to motivating the patient to achieve normal motion of the neck through normal eye-head movements.

Sample Exercises

The following exercises may be used in the middle to late phases of mobilization and strengthening and the development of daily activities:

1. PNF:

   - scapula pattern
   - exercise 1 under sample exercises for cervical vertebral fracture

2. Exercises with a ball:

   - rolling the ball forward, to the side, down to the feet, and so on
   - rolling the ball in different directions
   - lifting the ball with both hands and touching it down at different points
   - giving the ball with both hands to a partner, using different heights and distances
   - moving the ball away from the body in different patterns (figure-eight pattern, circle, waves, and so on)

3. Exercises with a towel:

   - wiping across a smooth area (for example, a door)
   - playing at window cleaning

During the exercises listed under 2 and 3, the patient is asked to continuously observe the motions of the equipment (eye-head movement).

FRACTURES OF THE THORACIC VERTEBRAE

Mechanism and Types

The section of the spine with thoracic vertebrae is characterized by its physiologic kyphosis and its support to the rib cage. Falls and, to a lesser extent, compression cause fractures of the vertebral body. These fractures are mostly flexion fractures resulting from hyperflexion, thus wedge-shaped deformations are frequently observed. Less frequent are hyperextension fractures. Compression fractures are observed with preexisting diseases (osteoporosis).

Medical Management

In treating fractures of the thoracic vertebrae, it must be remembered that the bony thorax can be elevated only to a limited degree and that the fracture stabilizes rapidly. This makes a long bed rest unnecessary and, in many cases, permits early functional treatment and the early application of three-point stabilization braces.

In the physical therapy, it is advantageous to treat not only the twelfth but also the two next higher thoracic vertebrae, following the procedures used for injuries of the lumbar vertebrae. The time period of the treatment depends on the extent of injury but lasts approximately 8 to 12 weeks.

Patients with unstable fractures have to be placed on a hard surface in a lordosing position, which can then be raised steadily. After four to six weeks, mobilization can be started with the patient in a back brace.

Physical Therapy

The following therapeutic regimen is applicable for fractures of the first nine thoracic vertebrae:

Symptoms

With the patient lying in a supine position, the following symptoms are observable:

- decreased circulation, especially in areas in contact with the bed
- limited motion and weakness of the extremities, caused by the supine position
- possibly decreased respiratory function
- pain at the fracture site

Therapeutic Aspects

- pressure sore prophylaxis: (1) position changes, (2) training in importance of position changes, (3) isometric and dynamic exercises
- pneumonia and thrombosis prophylaxis: (1) breathing exercises, (2) active exercises
- circulatory stimulation: isometric and active exercises
- mobilization during the consolidation period: (1) maintaining the free motility of all joints by active exercises, (2) active exercises of the extremities while lying on the side and in prone positions.
- mobilization after the consolidation: exercising the free mobility of the thoracic spine in different positions by active movements or using equipment

- strengthening: isometric and resistive exercises of the muscles of the extremities and trunk (especially extensors of the back) in different positions
- training in activities of daily living: (1) decreasing the limited motion by active movements, with or without equipment; (2) exercising activities of daily living

Course of Treatment

In the first week, therapy is implemented with the patient lying on the back. It begins with exercises to improve ventilation and to stimulate circulation.

The extremities are treated with dynamic, mobilization, and strengthening exercises. The fracture must remain immobilized during these exercises; the patient cannot change position himself or herself, and the shoulders and pelvis have to rest on the bed. Isometric contractions of the abdominal and thoracic muscles are allowed.

Beginning in about the second week, the patient learns to turn from the side to the prone position and, in that position, to strengthen the back muscles through resistive exercises to the cervical extensor muscles.

After the third week, exercises in the therapeutic pool (transported in supine position, and sliding into the water on a hydraulic lift) are incorporated in the therapeutic program. In deep water, the patient begins to stand up and walk. At first, the water level should be shoulder high; during the following days it should be lowered gradually. After one week, the patient is allowed to bear weight outside the water. To avoid chronic back pain, the patient is advised to sit as little as possible and to walk and lie down while maintaining good posture.

Now the mobilization phase of physical therapy can be started. The patient is encouraged to utilize the entire range of motion of the thoracic spine. Improvement of limited mobility and resumption of daily activities normally require a considerable period of time. These goals are achieved primarily through exercises on a mat, by working with equipment in various positions, and by repeatedly squatting and standing up in front of a mirror. Occupational and recreational therapy complete the rehabilitation program.

Sample Exercises

The following exercises may be used to strengthen the back muscles, mobilize the thoracic spine, and train for activities of daily living:

1. Strengthening of the back muscles (early to middle phase in bed, prone position):

- elevating and turning the head
- elevating head and shoulders from the bed
- alternately elevating extended arms
- elevating one arm and one leg diagonally from the bed
- with arms flexed in a U-position or hands underneath the forehead, elevating the head-shoulder-chest slightly from the bed

2. Mobilization of the thoracic spine (middle to late phase):

   - crawling position
     —flexion/extension of the spine in a "humped cat" exercise
     —flexion/extension of the spine by touching the knees to the forehead
     —lateral flexion of the spine, elevating the arm and swinging it alongside the head to the opposite side, or stretching with both hands to the side with the pelvis fixed
     —extension plus rotation of the spine by swinging the arms side to side
   - Squatting position
     —elevating a weight from the floor with both hands and stretching (arms overhead)
     —rolling a ball from side to side with both hands
     —rolling a ball from side to side with both hands but at the end of each movement taking one hand off and raising it across the head, bending the spine laterally
     —rolling a ball from side to side with both hands but from different positions in front of the body, then alternately stretching the right and left arm toward the ceiling in external rotation

3. Activities of daily living (late phase):

   - standing up and lying down
   - putting on shoes and stockings
   - picking up items from the floor
   - simulating the use of various household and working tools
   - getting in and out of the car

FRACTURES OF THE LUMBAR VERTEBRAE

Mechanism and Types

Vertebral fractures are usually observed in the area of the last thoracic and first lumbar vertebrae (T12 and L1). The wedge-shaped deformations are predominant;

they result primarily from flexion injuries and compression fractures during a fall on the back.

Clinical Considerations

Pain is felt most acutely at the site of the injury but also radiates to the thoracic spine. Sometimes a retroperitoneal hematoma can mimic acute abdominal trauma. Locally, a distinct gibbous across the transverse process in the paravertebral muscle area can be painful to palpation.

Diagnosis

Radiograms will usually show the lesion precisely. The lesion has to be differentiated from osteoporotic deformation and from juvenile growth disturbances (frequently a wedge shape of L1 in Scheuermann's disease).

Medical Management

Fractures of the transverse and horizontal processes do not usually require special treatment. Temporary immobilization for two to four days leads to quick pain relief. Fractures and slippage of L5-S1 (traumatic spondylolysis) involving the intervertebral disks are very rare. The treatment in such cases is the same as that for unstable vertebral fractures (see section on fractures of the thoracic vertebrae).

Physical Therapy

Fractures of the Tenth Thoracic Vertebra to the Fifth Lumbar Vertebra

Basically, the treatment is the same as that for fractures of the first to ninth thoracic vertebrae. However, the time period is different, since the patients are not allowed to bear weight before two weeks. If the fracture is at the level of the twelfth thoracic to the fourth lumbar vertebrae, the hip joint should initially be flexed minimally to avoid an unfavorable lordosis at the fracture site.

Fractures of the Transverse and Horizontal Processes

Exercises for mobilization, strengthening, and activities of daily living can be started immediately, since normally immobilization is not necessary. In the first days after the accident, the patient suffers considerable pain. The physical therapist must keep this in mind and schedule the progress of the exercises carefully.

Complications of Vertebral Fractures

The prognosis for spinal injuries is favorable if the spinal cord is not involved. In some cases, however, deformation of the vertebral bodies and limitations in the vertebral range of motion can persist. Insufficient healing of the simultaneously injured ligaments can result in unstable segments and problems in moving and weightbearing (instability symptoms). In a few cases, a spondylodesis may be noted.

LESIONS OF THE SPINAL CORD

Mechanism of Injury

Damage of the spinal cord is possible in severe spinal injuries with unstable vertebral fractures, and dislocation of fragments into the spinal canal, extensive hematomas, or subluxations of the vertebrae.

Types

The complications can vary in degree, from a slight loss of neurologic functions to a complete loss of function. The complications may involve impairment of:

- motor function
- sensory function
- reflexes
- autonomic nervous system

Diagnosis

The segmental level of the lesion is determined by the types of disturbances and losses in function, as noted above. Lesions at the level of C1 and C2 are often fatal. Lesions between C3 and T2 may cause Horner's syndrome and paralysis of the upper extremities. Lesions between the T12 and S2 can result in paralyzed lower extremities. Damaged third to fifth sacral nerves will interfere with the function of bowel and bladder control.

Medical Management

The initial treatment of patients with fractured vertebrae and neurologic deficits is very problematic. The patient must be transported to the hospital in a supine position and without the slightest vibration to avoid dislocation of the spinal segments and further lesions of the spinal cord.

Primary surgery is usually not indicated in completely paralyzed patients. However, if an incomplete paralysis is progressing, surgical decompression and immobilization are required. Stabilization of the fracture, even in completely paralyzed patients, improves the efforts of intensive care. The patient is placed in a rotating bed or on special pillows to avoid pressure areas. The urinary bladder has to be catheterized, and the intestinal function must be supported. Rehabilitation is then begun in a special spinal cord injury center.

Chapter 5

Thoracic Injuries

INJURIES TO THE RIB CAGE

Mechanism of Injury

Thoracic injuries to the rib cage are usually caused by a direct impact.

Types

Injuries of the rib cage are common, about 5 to 8 percent of all bone injuries. They are classified as blunt (closed) or perforating (open) chest trauma. Closed thoracic injuries can range from simple bruises, to sprains and fractures, to compressions with serious injuries of the internal organs. Injuries of the heart and lungs are also possible in perforating thoracic lesions. Medical management of closed thoracic injuries and the management of their related complications, such as hemothorax, pneumothorax, mediastinal emphysema, and injuries of the heart and its main vessels, is described in standard textbooks on internal and surgical medicine. In the present context, the discussion is confined to bony injuries of the thorax.

RIB FRACTURES

The rib can be fractured in any section between the dorsal to the ventral insertion. Usually, the injury occurs as a bending or disruption fracture or an impression fracture after a severe impact. An extensive impact might cause the fracture of several ribs or even bilateral rib fractures. With increasing age, the traumatic risk is increased because of the loss of elasticity of the rib cage. Bony lesions of the thorax are rare in children.

Clinical Considerations

The primary symptom of a thoracic injury is painful respiration. The site of the injury can be determined by crepitation, pain induced by palpation, and compression. Rib fractures are characterized by delayed inspiration on the involved side, and uneven spots along the rib can be palpated.

Diagnosis

The diagnosis is verified radiographically. However, rib lesions at the transition from the bony to the cartilaginous part will not show up in a radiogram. Additional radiograms in several planes may be required to evaluate the sternum. Also, it is recommended that the thoracic spine be included in the radiogram of a sternal fracture, since the combination of a sternal fracture and a comminuted fracture of a thoracic vertebra is not uncommon.

Medical Management

The medical management is aimed at maximum immobilization. This is accomplished with an elastic bandage, applied around the thorax in circular layers or with a "tile" pattern using adhesive tape. In elderly patients, the prevention of further complications, especially bronchopneumonia, is most important. In some cases, pain relief by paravertebral injections is indicated. Hospitalization may be necessary with multiple rib fractures and in the case of such complications as hemothorax, pneumothorax, or pneumonia. The unstable thorax is "splinted" internally by mechanical ventilation (positive pressure).

Physical Therapy

Physical therapy for rib fractures is the same as that after thoracic surgery, depending on the extent to which internal organs are involved. The most important treatment is respiratory therapy, which supports inspiratory and expiratory movements.

The following therapeutic regimen is applicable in fractures of single ribs and a stable thorax.

Symptoms

With the patient lying in bed with the chest elevated, the following symptoms are observable:

- possible hematoma and enlargement in the area of the fracture
- reduced thoracic expansion during respiration and while coughing, at rest, or moving around, with correspondingly reduced vital capacity
- pain during breathing, coughing, and moving
- tendency to avoid activities

Therapeutic Aspects

- pneumonia prophylaxis: (1) respiratory therapy, (2) active breathing exercises
- thrombosis prophylaxis and circulatory training: (1) respiratory therapy, (2) active movement, (3) stimulation to sit, stand, and walk
- circulatory support: (1) active movement, (2) therapeutic pool
- mobilization and strengthening: (1) active motion of the extremities and trunk plus extremities, (2) active stretching and turning positions of the trunk, (3) active motion against resistance and with equipment
- exercising through activities of daily living

Course of Treatment

Physical therapy starts on the day after the accident and involves primarily breathing exercises and mobilization of the patient. Breathing exercises help the patient include the injured side of the thorax in the respiratory movement, to intensify the respiration, and, if necessary, to cough up secretions.

The therapy involves the following exercises:

- inspiratory training with a spirometer
- inspiratory support through manual contact with the thorax
- combination of movement and breathing exercises initially lying in supine position on the normal side, later in stretching and turning positions, in a sitting up position
- combining vocal exercises and breathing

The movement of secretions and their removal by coughing is improved by vibration and exercises with humming and growling sounds.

The better the physical therapist supports the patient in coughing, by splinting the fracture site with his or her hands, the less severe the patient's pain will be. Also, the patient will be less anxious and cooperation with the therapist more successful. The therapist should show the patient at the beginning of the treatment how to provide self-support (''splinting'') when coughing up secretions.

Active movement of the arms is very important in the mobilization process. The patient is advised to utilize the full range of motion of the shoulder joint, with the arm on the injured side (inspiratory stimulus by stretching, prevention of secondary contractures). These movements are easily combined with inspiratory exercises. As early in the treatment as possible, the therapist should work with the patient in changing positions from one side to the other, in sitting up, and in moving around. The therapist should encourage the patient to exercise while alone, especially in standing and walking. As exercises in new positions and activities of daily living are introduced, they should be done both with and without equipment. Exercises should combine movement and breathing. Approximately one week after the accident, the patient is allowed to swim.

Sample Exercises

The following exercises may be used in middle and late phases of recovery for pneumonia prophylaxis, mobilization, and activities of daily living. In the first set of exercises, the patient is positioned side-lying, with legs slightly bent, the lower hand underneath the head and the upper hand loosely in front of the body. The therapist's hands are placed on the sides of the patient's thorax. The patient breathes against the hand contact while inhaling with one nostril pressed closed or by "sniffing." The exhalation is combined with humming sounds.

The following exercises can also be performed:

- during inspiration, stretching the upper foot out into eversion and plantarflexion (the therapist maintains contact with the heel and/or thorax), then making expiration sounds while relaxing the stretched foot
- during inspiration, swinging the upper arm across the head and then pushing it toward the floor, then making expiration sounds while relaxing the stretched arm and hand
- combining the previous two exercises as one
- swinging the upper hand from extension-adduction-internal rotation in front of the knees into flexion-abduction-external rotation next to the ear (one hand of the therapist at the patient's pelvis, the other one leading the patient's hand with a slight resistance), while combining the first part of the exercise with inspiration and the second part with expiration
- performing the last exercise without the second movement; instead, relaxing at the upper end of the movement with expiration; during the next inspiration, reaching the arm toward the floor and relaxing during audible expiration; next with the knees flexed and pressed tightly against the bed while alternating tension/inspiration and relaxation/expiration, or by closing the fist, expiring, then opening the fist

- working with a tennis ball or small dumbbell while in a side-lying position; using the equipment in a diagonal upper extremity motion or bringing it from a stretched position in a semicircle to the front of the knees and back again, upward into the stretched position (the stretching-extending phase combined with inspiration, the relaxation phase combined with an audible expiration)

The following stimulating exercises may be used for activities of daily living:

- combing the hair
- tying a scarf on the head
- hanging wash on high line
- changing a lightbulb
- holding onto an upper railing in a bus (combining movement and respiration)

MULTIPLE RIB FRACTURES

Multiple rib fractures pose risks of hemo- and pneumothorax, collapsed lungs, and pneumonia. Therefore, in the first days after the accident, the patient is observed during hospitalization.

The design of physical therapy is considerably more complicated than with a simple rib fracture. Preventive measures against pneumonia and thrombosis are combined with positioning, isometric exercises involving the muscles adjacent to the endangered area, and active movements.

Mobilization may be delayed while the primary care by respiratory therapy is given. This may involve intensified procedures, dissolving secretions, and supporting expectoration, while the patient remains in a supine position for a period of time. The dosage and intensity of the therapeutic program is dependent on the individual situation and must be coordinated with the physician.

STERNAL FRACTURES

Mechanism of Injury

A sternal fracture is usually caused by a direct impact on the anterior thoracic wall. Less frequently, it occurs in connection with a fracture of a thoracic vertebra by a "jackknife" flexing mechanism. It is characterized by local pain and sometimes by the palpation of a bony ridge.

Types

In a simple fracture, the pain and functional limitation is inhibited respiration and is relatively minor. Radiographic examination is needed to verify the diagnosis, and it should be combined with a radiogram of the thoracic spine. An unstable sternal fracture, especially in connection with fractured ribs, causes respiratory disturbances, such as collapsing thoracic wall during inspiration, resulting in hypoxia.

Medical Management

Undisplaced fractures do not usually require treatment, other than the use of an analgesic in some cases. Displaced fractures with a complete separation of the fragments and possible accompanying rib fractures are stabilized by internal pneumatic ''splinting'' of the thoracic wall, for example, with intubation and positive pressure artificial respiration.

Physical Therapy

Physical therapy for a sternal fracture is the same as that for fractured ribs, except that mobilization of the trunk (stretching and turning) has to be started later and more carefully. Usually, sternal fractures heal within 6 to 8 weeks without complications.

Chapter 6

Pelvic Injuries

MECHANISM OF INJURY

Pelvic injuries are often caused by a severe impact and occur together with lesions of the internal pelvic organs. The accompanying injuries, caused by indirectly transmitted forces (acetabular fracture or muscles torn off from their insertion), are often the focus of medical attention because of their urgency.

PELVIC FRACTURES

Types

Pelvic fractures are classified as fractures of the ilium, superior or inferior pubic ramus, ruptures of iliosacral joint and pubic symphyses, and fractures of the pelvic ring (see Figures 6–1 and 6–2). Fractures of the sacrum and ischium also occur. Isolated iliac fractures are of little clinical significance. However, fractures of the

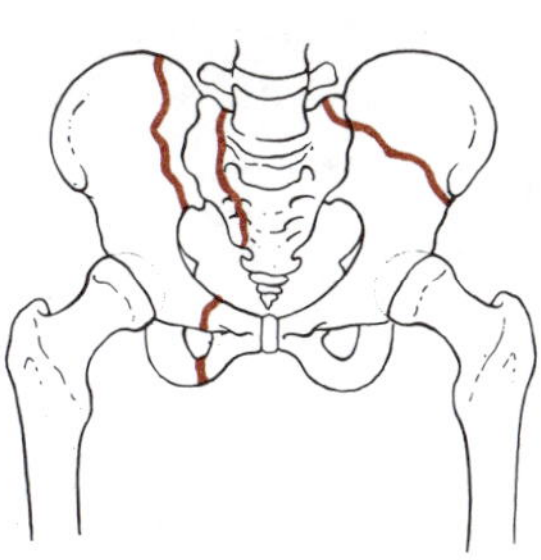

Figure 6–1 Types of pelvic fractures. Left: ring fractures. Right: iliac fracture

Source: Schlosser and Kuner, 1980

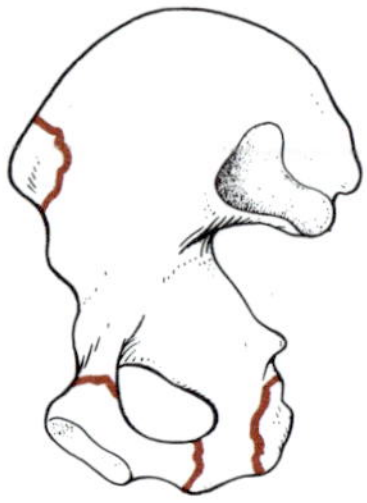

Figure 6–2 Isolated pelvic fractures. Dislocated fracture of the anterior superior iliac spine and the ischium (especially in children) and solitary fractures of the pubis and ischium (especially in older patients)

Source: Schlosser and Kuner, 1980

pubic rami, pelvic ring, and ruptures of the symphysis and the iliosacral joint are clinically important. Fractures of the pelvic ring in particular interfere with the stability of the pelvis, which supports pressure between the trunk and lower extremities. The pelvic ring is formed by the sacrum and the pelvic bones (bilateral ilium, ischium, and pubis), and it is connected by pseudojoints or symphysis at the iliosacral and pubic junctures. The tight ligaments and strong muscles create a firm but elastic component for the pelvis.

These soft-tissue structures are often involved in severe pelvic injuries caused by direct impact. Simultaneous injuries are typically found to the urinary bladder, urethra, genital organs, and sciatic nerves. All pelvic fractures can cause major hemorrhages. Accompanying contusion of fat and muscle tissue creates the risk of fat emboli.

Clinical Considerations

The diagnosis of a pelvic fracture might be missed at first because of multiple trauma. The predominant symptom is a compression pain of the pelvis in the sagittal and frontal planes and pain with compression of the extended leg. Extensive hematoma formation may also indicate a pelvic injury, which can be diagnosed radiographically.

Medical Management

The healing of a fractured pelvic crest takes several weeks; in pelvic ring fractures with displacement of the fragments, it might take months. Fractures of the ilium and the pubic rami with remaining pelvic stability usually do not require immobilization. Fractures of the pelvic ring without major dislocations require bed rest for up to four weeks. Dislocated fractures and ruptured iliosacral joints and symphyses usually require surgery, especially if simultaneous injuries have to be operated on. Nonsurgically, bed rest and special suspended positions can be used.

First-aid manipulations at the site of the accident and at admission to the hospital include shock prophylaxis treatment; monitoring of blood pressure, pulse, and urinary output; and prophylactic measures against fat embolism and for the control of bowel and bladder functions. Treatment of severe accompanying lesions, such as ruptures of the urethra and urinary bladder, are particularly important.

Physical Therapy for Fractures of Single Pelvic Bones Without Dislocation

The following therapeutic program is applicable.

Symptoms

With the patient lying on the back, the following symptoms are observable:

- decreased circulation, especially in areas against the bed
- possible hematoma and swelling at the fracture site
- decreased range of motion of the legs, with accompanying pain and general inactivity
- pain around the fracture site

Therapeutic Aspects

- pressure area prophylaxis: (1) positioning, (2) isometric and dynamic exercises, (3) training to change positions
- pneumonia and thrombosis prophylaxis: (1) breathing exercises, (2) isometric and dynamic exercises, (3) changing positions
- circulatory stimulation: (1) isometric and dynamic exercises, (2) changing positions, (3) therapeutic pool
- mobilization: (1) isometric and dynamic exercises, (2) position changes plus active movement in each position, (3) therapeutic pool
- strengthening: (1) dynamic muscle work against resistance with and without equipment, (3) therapeutic pool
- ambulation

Course of Treatment

On day after the accident, physical therapy starts with prophylactic treatment. In addition, with the patient lying on the back, the legs are moved passively and actively in all directions without resistance. It is recommended that support be used with these exercises during the first days.

Beginning in about the second week, the physical therapist trains the patient to turn to the normal side and to the prone position. After the new position is achieved, the patient is instructed in mobilizing and strengthening exercises. Extension of the hip and the strengthening of the gluteal muscles are particularly important. At this stage, the patient is allowed to go into the therapeutic pool.

In the third week, weightbearing is usually performed in the pool. With the water at the level of the neck, the patient is allowed to walk (forward, backward, to the side). After another three days, weightbearing is tried outside the water. Walking aids such as walkers, crutches, or canes are provided, depending on individual needs, and then gradually withdrawn. Simultaneously with exercises to correct standing posture and ambulation, further strengthening exercises are done. As soon as the fracture is completely consolidated, resistive exercise, with and without equipment, can be applied, using PNF, mat activities (using all-fours positions, with the patient sitting on the heels, kneeling, and so on), resistive equipment, and the exercise bike. In addition, swimming, occupational therapy, and recreational therapy can be prescribed.

Sample Exercises

The following exercises may be used in the middle phase of mobilization and strengthening and in walking exercises in the therapeutic pool. The first set of exercises is performed with a low water level, with the patient lying in a supine position (possibly with a ring around the neck) and holding onto a rail:

- flexion-extension, abduction-adduction, internal-external rotation of the knee and hip, if possible, doing all these exercises bilaterally
- PNF patterns: from extension-abduction-internal rotation into flexion-adduction-external rotation
- flexing the leg and alternately bringing it toward the right and left armpit and stretching it out again
- dipping flexed knees from side to side into the water
- pushing extended legs from side to side through the water

The following set of exercises is performed with a low water level, with the patient in the prone position and holding onto a rail:

- contracting the gluteal muscles and lifting the heels out of the water
- dipping the extended legs into the water and pushing them back up (heels out of the water)

- same as above, but moving the legs opposite each other (one up, one down)
- abduction-adduction, with fully extended hip
- pulling both extended legs from side to side
- from full extension, pulling both legs into a complete hip-knee flexion and back into extension
- same as above, but pulling both knees toward the right (left) shoulder
- bending both closed legs, extending them downward, letting them float to the surface in extended position, then stretching again (pushing the heels out of the water)

While sitting at the side, the patient ''bicycles'' backward and forward. Finally, while standing on the pool floor, with the water level adjusted as necessary, the patient can practice weightbearing while walking forward, backward, then to the left and right sides.

Physical Therapy for Nonsurgical Treatment of a Fracture of the Pelvic Ring

The following therapeutic program is applicable.

Symptoms

With the patient either in a supine position or, in case of a dislocation, in extension traction and flat on the back, the following symptoms are observable:

- possible hematoma in the pelvic area
- decreased activity of the legs and decreased general activity because of pain
- decreased blood flow in all areas against the bed
- possibly decreased function of respiration and circulation

Therapeutic Aspects

The treatment of a fracture of the pelvic ring is the same as that for fractures of single pelvic bones without displacement, except that the exercises (especially active movements, position changes, PNF) are begun later and increased in intensity more gradually.

Course of Treatment

In the first week, the most important therapeutic procedures are prophylactic measures to aid bedridden patients. Usually, at this stage isometric exercises are indicated for both legs. In the following week, dynamic exercises are added. These are initially done with assistance.

In about thc fourth week, pool therapy is also started. The patient is allowed to exercise in all sitting and lying positions at the side of the pool with the water level adjusted to the neck level.

Beginning in about the sixth week, exercises while prone and, to the extent permitted by radiogram and clinical findings, exercises in a side-lying position can begin.

Beginning in the eighth week, the patient is usually allowed to stand and walk in the pool, with the water level just covering the shoulders. A few days later, the weightbearing phase starts outside the water, first with a walker, later supported by crutches. The patient must wear elastic stockings because of the preceding long period of bed rest. All aids are gradually withdrawn during the following weeks, at a rate suited to the individual situation.

In addition to participation in walking exercises, the patient also engages in intensive exercises in crawling, kneeling, and standing positions. Resistive exercises, without restrictions, including PNF, the exercise bike, and recreational therapy, are allowed to achieve further improvement of function and to motivate the patient to persevere after the consolidation of the fracture.

In patients in traction, the treatment is basically the same, but it develops more slowly. The patient engages in weightbearing exercises after 12 weeks.

The treatment of patients with surgically repaired fractures of the pelvis ring is little different from that used in nonsurgical treatments. However, dynamic exercises in all supported positions is allowed earlier, given sufficient bony stability exists, and usually the patient does not begin weightbearing before three months.

Sample Exercises

The following exercises may be used in the late phase of mobilization and strengthening when weightbearing is allowed. In the crawling position, the following exercises may be performed:

- holding against resistance to the pelvis, shoulder, head, and extremities, to each body part at a time, and to two parts at a time
- flexing the head and knees toward each other beneath the abdomen
- extending one leg straight back while maintaining the toes on the supporting surface; doing the same exercise with hip abduction
- combining both of the previous exercises in one continuous movement
- extending one leg and elevating it with the ankle plantarflexed until full extension of the hip is achieved

- combining full flexion (head and knee flexion exercise) with full extension (previous exercise)
- elevating one arm and the opposite leg

The following exercises are performed with the patient sitting on the heels (the patient can also hold on to a rail):

- moving shoulder and pelvis toward each other
- combining resistance against the pelvis with the end phase of the diagonal arm position; in extension-abduction-internal rotation or flexion-adduction-external rotation; if necessary, resisting the arm to improve the pelvic motion
- getting up into a kneeling position against the therapist's resistive contact on the pelvis

The first two exercises described above may also be done in the kneeling position.

ACETABULAR FRACTURES

Mechanism of Injury

Acetabular fractures, with or without subluxation of the femoral head, are caused by a strong direct force transmitted to the femoral head and neck. Usually, these fractures occur in a car collision by an impact against the femur flexed in the hip joint.

Types

In relation to simple fissures, a fracture of the acetabulum and its depression inside the pelvic cavity is more serious clinically. Such a fracture is called a central subluxation of the hip.

Medical Management

Uncomplicated fractures are treated nonsurgically. Weightbearing on the joint must be avoided for 8 to 12 weeks. Frequently, dislocated fractures have to undergo surgery to reduce and internally stabilize the fragments in order to restore the congruence of the joint. Massive comminuted fractures are treated with an extension splint and traction.

Complications

With an acetabular fracture, a common risk is the possibility of necrosis of the femoral head. This may also occur after a fracture of the femoral neck. A post-traumatic deforming arthrosis is frequently observed. A temporary preventive measure is the application of a supporting device. A long-term prognosis is questionable.

Physical Therapy

The following therapeutic regimen is applicable to nonsurgically treated central subluxations of the hip.

Symptoms

With the patient in a supine position, the injured leg splinted in extension and axial traction of approximately 1/10 the body weight for 6 to 8 weeks, side pressure of 1.5 kg for about six weeks, the following symptoms are observable:

- possible hematoma and swelling around the fracture site
- mobility generally limited by pain and traction
- decreased blood flow from position of injured leg in extension
- decreased blood flow in all areas against the bed
- changed turgor of the soft tissue
- during the time of necessary immobilization, increased inactivity atrophy of the injured extremity, especially the quadriceps and gluteal muscles, together with the expected limited range of motion of the knee and hip joint (which is initially unverifiable)
- sometimes decreased respiratory activity

Therapeutic Aspects

- pressure area prophylaxis: (1) positioning, (2) isometric and dynamic exercises, (3) training in changing positions
- pneumonia and thrombosis prophylaxis: (1) respiratory therapy, (2) isometric and dynamic exercises
- circulatory stimulation: (1) isometric and dynamic exercises, (2) training in changing positions, (3) use of therapeutic pool
- mobilization: (1) isometric and dynamic exercises, (2) training in changing positions and active motion in each position, (3) possibly assistive dynamic

splint (Frankfurt model), (4) relaxation techniques with ice packs; (5) movement in the therapeutic pool

- strengthening: (1) dynamic exercises with and without equipment (resistive material, exercise bike), (2) PNF, (3) posture training, (4) movement in the therapeutic pool, (5) recreation therapy
- training in activities of daily living: (1) walking exercises, (2) possibly activities such as sitting down, standing up, picking up things, climbing onto a footstool, and so on.

Course of Treatment

In the first four weeks, the therapist works with the patient using the guidelines listed in the physical therapy section for fractures of single pelvic bones without dislocation. In addition, the toes and foot of the injured leg can be moved actively, without limitations, and isometric muscle contractions may be done carefully with intensive contraction of the gluteal muscles to avoid a flexion contracture of the hip joint.

In the sixth week, at the earliest, the lateral traction can be removed. The therapist can also remove the axial extension weight during the exercises and work on supported flexion of the knee and hip joint.

Beginning in about the eighth week, the extension splint is no longer necessary. The leg can be laid beside the splint. In this position, full flexion and extension of the knee and hip joint are attempted. In case of a flexion contracture of the hip, the compensatory lordosis of the spine (which will occur repeatedly during the course of the exercises) must be corrected by flexing the head and/or adducting the normal leg. Mobilization of the knee joint is especially difficult because of the extension traction. The therapist must work very carefully in this early phase, but often even minor modifications will be successful, for example, through contract-relax at the end of the movement and repeating to gain more flexion. It is recommended that the continuing exercises be supported with ice packs.

If the therapy is progressing well, the splint can be removed for progressively increasing time increments after the tenth week, or it can be replaced by an assistive dynamic splint (see Figure 8–2). Now the patient is progressively mobilized and strengthened in supine, side-lying, and prone positions. The prone position is best for the mobilization of an extension contracture of the knee joint, as well as for gaining full extension of the hip and the strengthening of the gluteus maximus muscle. Hip abduction has to be watched very carefully to avoid an unfavorable stress on the fracture site. Simultaneously with this exercise program, the patient participates in exercises in the therapeutic pool.

When the fracture is sufficiently consolidated (at the earliest, after about 12 weeks), the patient can begin to walk in deep water. A few days later, the patient can begin walking on land and using elastic stockings. At first, the patient uses a walker, and then the aids (walker, crutches) are gradually withdrawn, depending on the consolidation of the fracture and muscle strength. Unassisted walking, even after a complete consolidation of the fracture, is allowed only after the muscles are built up sufficiently so there is no tendency to limp and a Trendelenberg is not present.

Because of the long immobilization period, it is necessary to concentrate intensively strengthening and endurance training on the mat with the patient in all positions, from lying down to standing up. The rehabilitation program concludes with use of an exercise bike, occupational therapy, and recreational therapy.

After surgery for a central hip subluxation there are two treatment possibilities following the open reduction and internal fixation: either the acetabulum has to be relieved for a certain period of time by traction or it may be determined that it is not necessary. In both cases, the physical therapist employs the same methods as for nonsurgical management. For about three months, the patient is mobilized and strengthened without weightbearing. However, the dynamic exercises can be started earlier than in the nonsurgical treatment. Weightbearing cannot be begun earlier than 12 weeks after surgery.

Sample Exercises

The following exercises may be used to support blood circulation, mobilization, and strengthening during the early phase, with the patient lying on the back:

- moving the toes: flexing, spreading, extending
- moving the foot: flexion, extension, inversion, eversion, circling
- isometric contraction of the quadriceps muscle including the foot
- isometric exercises of the entire leg
- working on correctly aligned flexion-extension of the knee and hip, without weightbearing; then using therapist support, with the distal forearm under the patient's hip and proximal hand under the patient's thigh (including the foot in the exercise)
- repeating the previous isometric contractions between the flexion/extension exercises

The above exercises may also be done using the assistive dynamic splint (see Figure 8–2).

Finally, these same exercises may be done dynamically, with weightbearing.

Periods of a few days to several weeks may be interposed between the three sets of exercises described above, depending on the individual situation.

TRAUMATIC SUBLUXATION OF THE HIP JOINT

Mechanism of Injury

Like serious pelvic fractures, a traumatic subluxation of the hip is caused by a strong, usually direct, impact with simultaneous rotary forces. This type of injury is increasing as a result of the growing number of severe car and workplace accidents. In car collisions, the knee joint hits the dashboard, resulting in a contusion of the thighs; the femoral heads are pushed out of the acetabulum, backward and upward, with a fracture and dislocation of an acetabular fragment.

Types

There are four types of subluxation, depending on the direction of the impact: backward and upward, backward and downward, frequently, subluxation forward and upward, and subluxation forward and downward (see Figure 6–3).

The subluxation can occur with or without an acetabular fracture. Thus the posterior subluxation may be further classified:

- first degree: posterior subluxation with the acetabulum intact
- second degree: with fractured acetabular rim, but with the femoral head inside the acetabulum after reduction
- third degree: posterior subluxation with fractured acetabulum and instability after reduction
- fourth degree: posterior subluxation with fracture of the femoral head or neck

Clinical Considerations

The frequently occurring posterior subluxation is characterized by an elastic fixation of the leg within the acetabular cup in an internally rotated and flexed position. In a posterior subluxation, the leg is rotated externally with a possible constriction of the arterial blood flow. The integrity of the femoral and the sciatic nerve must be checked. Radiographs reveal the type of subluxation. A pelvic radiograph must be examined carefully with regard to the width of the intra-articular space (narrowed or nonexistent) because a posterior subluxation is often missed.

Medical Management

The treatment of a subluxation consists of a quick reposition under general anesthesia. Immobilization is necessary for approximately two weeks. In third- and fourth-degree subluxations, surgical fixation of the fracture is indicated.

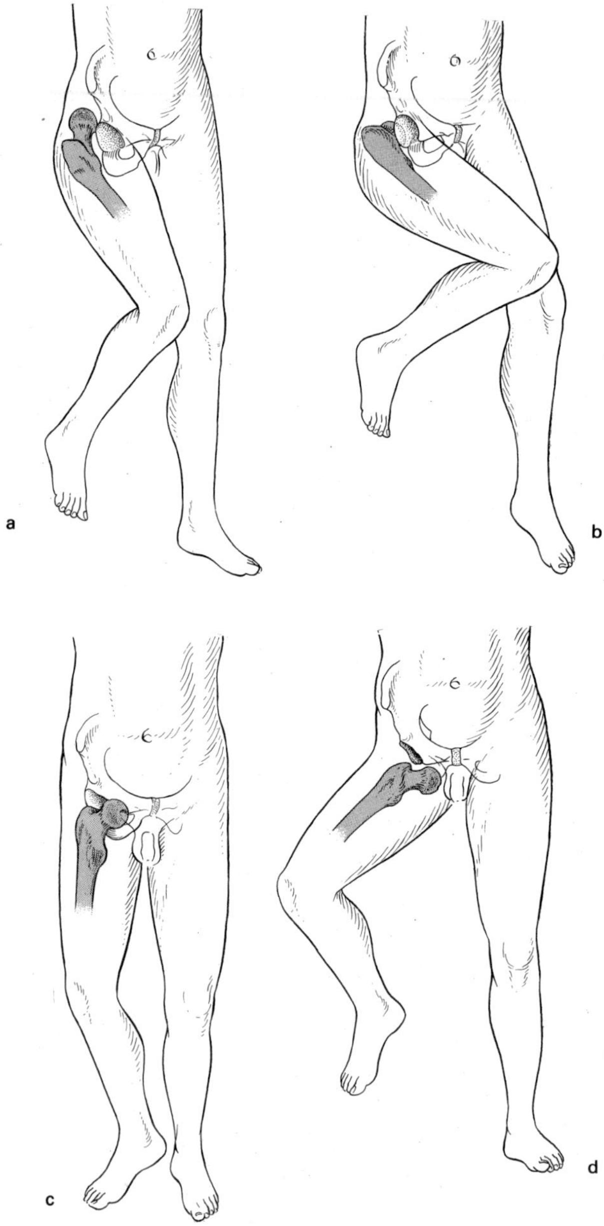

Figure 6–3 Types of subluxation of the hip joint a. posterior subluxation, b. posterior and downward subluxation, c. anterior subluxation, d. anterior and downward subluxation

Source: Schlosser and Kuner, 1980

Acute and Late Complications

Nerve lesions and disturbances of the blood circulation can occur in various degrees. If the vascular supply to the femoral head is disrupted as a result of subluxation (10 to 20 percent of the cases), a post-traumatic necrosis may occur. Post-traumatic arthroses are common if the acetabulum is involved. In some cases, extensive periarticular calcifications can occur.

Physical Therapy

The following therapeutic regimen is applicable in cases of traumatic subluxation of the hip without bony lesions.

Symptoms

With the patient in the supine position, the injured leg splinted, possibly in extension traction (1/10 body weight for ten weeks), the following symptoms are observable:

- possible hematoma and swelling around the fracture site
- all other symptoms match those appearing in a nonsurgically treated central subluxation of the hip

Therapeutic Aspects

The therapeutic procedures are basically the same as those for nonsurgically treated central subluxations of the hip. However, because of a shorter immobilization period, therapy is aimed less at an improvement of the contracture and more at strengthening the joint.

Course of Treatment

In the first days after the accident the therapist works with the patient, following the previously described guidelines for prophylactic therapy for bedridden patients. After about six days, careful movement in the supine position will be added.

After the third week, isometric exercises can be provided, and lying on the normal side and in the prone position can be started. Also, movement in the pool may begin.

Beginning in the fourth week, walking exercises can begin in the pool. At first, the patient is placed in deep water up to the shoulders. The water level is then lowered gradually. In the fifth week, at the latest, weightbearing can begin outside the pool. The transition is quickly made from a walker to a three-point gait pattern with crutches. The physician determines the amount of weightbearing and its increase so eventually the transition is made to a four-point gait pattern or walking with a cane. During the intensive correction of walking, the patient is also involved in strengthening exercises, posture training, recreational therapy, and occupational therapy.

Sample Exercises

The following exercises may be performed in the late phase of the mobilization, especially during strengthening, using a gymnastic ball:

- trying to sit on the ball against resistance
- from a sitting position on the ball, changing into a side position (normal side), with one or both arms supporting, balancing, holding against resistance
- getting on the ball by turning from a side into a prone position (both legs extended, arms supporting as necessary), then balancing and holding against resistance
- adopting a prone position on the ball, supported by extended arms, feet on the ground; then moving the ball and knee toward the arm, getting on the ball while sitting on the heels, pushing back again into an extended position, and maintaining the legs into an elevated position while balancing and holding against resistance

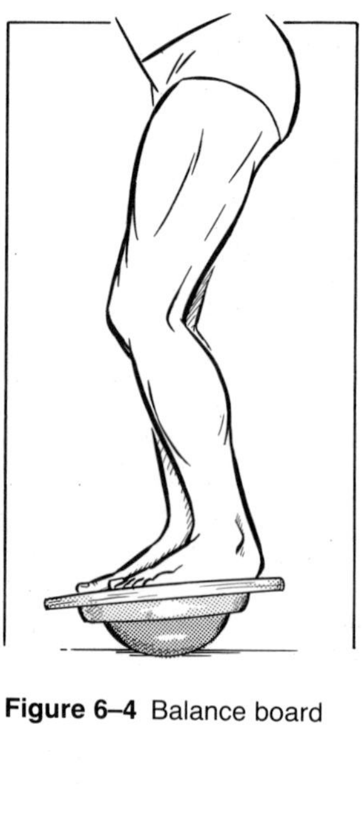

Figure 6–4 Balance board

In the following exercises, a balance board is used (see Figure 6–4); the patient may hold on to a rail, as necessary:

- balancing, stabilizing
- moving forward and backward, to the right and left; moving diagonally, holding, then moving again
- rotating, holding, then rotating again

Chapter 7

Injuries of the Lower Extremities

PROXIMAL FEMORAL FRACTURES

Mechanism of Injury

Fractures of the femoral neck are usually caused by a direct impact, especially by a fall on the hip (trochanter major) or leg. Older people with osteoporotic bones are particuarly at risk. Such people can also suffer from spontaneous fractures. Depending on the mechanism of the accident, bending and compression fractures and also shearing and torsion fractures, or a combination thereof, may be observed.

Types

Fractures of the femoral neck are classified as medial or lateral. Within these classifications there are intermediate types. Medial fractures are either abduction (10 percent of cases) or adduction types (90 percent). In an abduction fracture, the femoral head is pushed against the cranial part of the neck fragment, resulting in an enlarged angle of the femoral neck (valgus position) with limited stability. The adduction fracture results in a varus position: the fracture gap is widened proximally and the femoral neck is shifted proximally toward the femoral head. This type of fracture is unstable.

Pauwels' classifications of three types of fractures of the femoral neck are useful in diagnosis (see Figure 7–1).

- Type I: The fracture results in displacement of 30 degrees or less with the femur in abduction. The prognosis for a complete healing is favorable and the risk of necrosis is low.

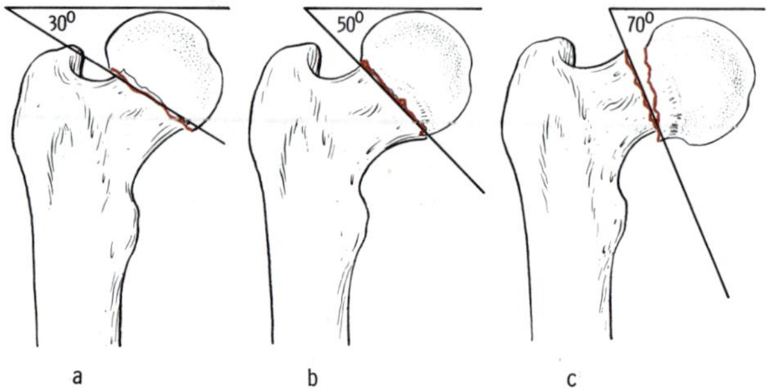

Figure 7–1 Fractures of the proximal femoral neck: a. abduction type (I); b. adduction type (II); c. adduction type with separation (III)

Source: Schlosser and Kuner, 1980

- Type II: The fracture results in displacement of 50 degrees with the femur in adduction. The healing tendency is satisfactory and the risk of necrosis is moderate.
- Type III: The fracture results in displacement of 70 degrees with the femur, resulting in a 20 to 30 percent risk of pseudarthrosis formation since the biomechanics for good realignment are most unfavorable.

Proximal fractures of the femoral neck occur intracapsularly and are accompanied by a hematoma. This results in increased pressure in the joint space and consequently in a disturbed vascular supply to the femoral head, which increases the risk of necrosis of the femoral head.

Distal fractures of the femoral neck are distinguished from proximal fractures (see Figure 7–2), since distal fractures occur extracapsularly. Trochanteric and subtrochanteric femoral fractures occur below the femoral neck.

Clinical Considerations

The symptoms of a fracture of the femoral neck are pain while moving the hip joint or from pressure originating in the trochanteric area or sole of the foot. The leg may be shortened and rotated into a malalignment. Frequently, active motion of the leg within the hip joint is completely obstructed.

Diagnosis

In addition to a determination of typical clinical symptoms, a radiographic examination in two planes permits an exact classification of the fracture.

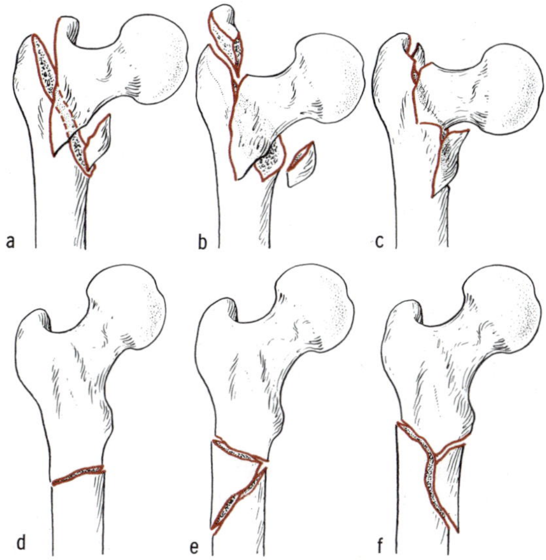

Figure 7–2 Intertrochanteric and subtrochanteric femoral fractures: a–c. intertrochanteric fractures; d–f. subtrochanteric fractures

Source: Schlosser and Kuner, 1980

Medical Management

The medical management is dependent on the type of fracture and the age of the patient. A comminuted fracture is generally treated nonsurgically. A Type II fracture of the femoral neck and fractures in the trochanteric area are usually treated surgically. In children, a proximal fracture of the femoral neck requires emergency surgical attention. The necessary steps are incision of the joint capsule, evacuation of the hematoma, and exact reduction and fixation with screws. Pauwels Type II, distal femoral neck, inter- and subtrochanteric fractures are stabilized by osteosynthesis up to 60 to 65 years of age (using intramedullary Ender or Kuntscher rods or plating with an angled plate). It is also necessary to determine the valgus position of the proximal segment. Pauwels Type II fractures require an additional osteotomy (excision of a lateral intertrochanteric wedge) to correct a valgus position, thereby decreasing the shearing forces on the fracture site and improving the pressure on the fracture fragments.

In patients over 60 to 65 years of age, especially those in poor physical condition, an endoprosthetic replacement (total hip prosthesis) may be required. In very old patients, only the femoral head is replaced, but normally both the head and the acetabular component are implanted (total endoprosthesis). This makes possible an immediate mobilization of the patient. Fractures stabilized by osteosynthesis require a lengthy period of immobilization.

Acute and Late Complications

In older patients cardiopulmonary complications may be life threatening. Local problems are necroses of the femoral head (15 to 30 percent of cases), pseudarthroses resulting from unfavorable biomechanical conditions (Pauwels Type III), and reaction to osteosynthetic materials. Arthritic deformations accelerate after a partial necrosis of the femoral head.

Physical Therapy for Nonsurgically Treated Fractures of the Femoral Neck

The following therapeutic program is applicable to nonsurgically treated fractures of the proximal and distal femoral neck.

Symptoms

With the fractured leg tending toward external rotation and placed between sandbags in slight abduction, the following symptoms are observable:

- decreased general activity caused by position and pain, possibly accompanying decrease of respiratory and circulatory function
- decreased blood flow in the leg and foot of the injured side and in areas against the bed
- possible hematoma and swelling around the fracture site
- skin turgor changes
- during the immobilization phase, onset of atrophy of all leg muscles
- limited range of motion, especially of the hip, but frequently also the knee joint (may not be initially noticeable)
- possible shortening of the leg (measurable at first only by tape with the patient in a nonweightbearing position, later in a weightbearing position)
- pain around the fracture site

Therapeutic Aspects

- pressure area prophylaxis: (1) isometric and dynamic exercises, (2) changing positions
- pneumonia and thrombosis prophylaxis: (1) respiratory therapy, (2) isometric and dynamic exercises

- circulatory stimulation: (1) respiratory therapy, (2) isometric and dynamic exercises, (3) changing positions, (4) therapeutic pool
- edema prophylaxis: (1) effleurage massage, intermittent pressure, (2) isometric muscle contractions, (3) active movements, (4) use of elastic stockings
- mobilization: (1) isometric and dynamic exercises, (2) relaxation techniques, possibly with ice packs, (3) changing positions, (4) use of therapeutic pool
- strengthening: (1) dynamic muscle work against resistance with equipment or PNF, (2) use of therapeutic pool, (3) recreational therapy
- training in activities of daily living: (1) walking exercises, (2) exercising in daily activities

Course of Treatment

In recent years, nonsurgical treatment of fractures of the proximal femur has been rare. When nonsurgical treatment is selected, the injured leg is placed between sandbags in a slightly abducted position. In the first one or two weeks, the patient exercises following the previously described guidelines for prophylactic therapy of bedridden patients. In pressure area prophylaxis, the patient learns, among other things, how to relieve pressure on the sacrum and buttocks (every hour if possible) by tensing the gluteal muscles and by elevating the body from the bed while supporting it with the normal leg and both arms.

After about the middle of the second week, isometric muscle contractions for the entire leg are added to the program. A few days later, dynamic exercises are allowed in the supine position, initially without weightbearing. At first, training in adduction and hyperextension should be avoided, because during the adduction an unwanted shearing force is transmitted to the fracture and in hyperextension the fracture site is unfavorably loaded by the long arm of the bony lever and pull of the iliopsoas and rectus femoris muscles.

Beginning in the fourth week, exercises in the side-lying position are possible. The patient lies on the normal leg with the leg slightly flexed and the therapist supports the injured leg while preventing adduction.

After about the sixth week, exercises are possible in the prone position. Exercises are aimed especially at strengthening in knee flexion, hip extension, trunk, shoulders, and arms. Beginning in about the eighth week, mobilization and strengthening in the swimming pool are also possible. However, swimming is not recommended for the first nine to ten weeks.

In the following period, the patient can sit on the edge of the bed more often and can also move about in a wheelchair. In case of a flexion contracture of the hip joint, however, this type of sitting position should not last longer than a half hour to an hour; the patient should be advised of the importance of compensating by lying in a prone position and exercising the hip extensor muscles.

In a normal course of treatment, walking exercises are started in deep water during the twelfth week. After the thirteenth week, the patient can begin ambulation outside the pool using elastic stockings and 10 kg of weightbearing with a walker. The walker is then gradually replaced by crutches. Corresponding to the rate of increased weightbearing, the walking aids are gradually withdrawn. Intensive walking exercises and postural training are mandatory.

In both physical therapy or occupational therapy, training in activities of daily living include such tasks as putting on stockings and shoes, sitting down and standing up, picking things up from the floor, and so on.

Sample Exercises

The following exercises may be used in the early phase of mobilization and strengthening:

- resistive exercises to the upper extremities and normal lower extremity
- scapula PNF pattern in prone position
- bilateral resistance to the latissimus muscles, plus trunk extension in the prone position

In treatment of a knee extension contracture in a prone position, the pelvis is fixed lightly by a belt and the posterior knee-thigh region may be treated with ice. Slowly the knee is flexed; then the flexor muscles are strengthened by repeated contraction (the tension and resistance must be gauged in such a way that any force caused by them does not create unfavorable stress or leverage at the fracture site).

The following exercises can be used to strengthen the extension of the hip in the prone position:

- isometric exercise of the hip extensor muscles directly or indirectly by elevating the head and upper body
- elevating the extended leg (partially supported) from the bed
- elevating the leg with the knee flexed from the bed
- elevating the flexed leg, extending it in that position, then laying it back on the bed in an extended position
- elevating the extended leg and the opposite extended arm at the same time, holding and then putting them back down
- mainly for younger, well-trained patients without a tendency to a lordosis: elevating both stretched legs, possibly elevating both stretched legs and arms at the same time, then slowly putting them down again

Physical Therapy for Surgically Treated Fractures of the Femoral Neck

The following therapeutic program is applicable to surgically treated fractures of the proximal and distal femoral neck, with the operated leg immobilized between sandbags in slight abduction.

Symptoms

- postoperative hematoma
- activity reduced by pain
- decreased blood flow in all areas against the bed
- possibly general, decreased respiratory function and circulation; and local indurated tissue, elevated temperature, contractures, and muscle atrophy

Therapeutic Aspects

The therapeutic procedures are the same as those for nonsurgically treated fractures of the proximal and distal femoral neck. For indurated heated scars, the application of ice packs supplements the indicated edema prophylaxis and stimulation of blood flow. A synthetic, but not wetting, ice material should be used until the wound is healed.

Course of Treatment

On the day after surgery, the physical therapist starts the prophylactic treatment for the bedridden patient, described earlier. In addition, isometric exercises are started with the injured leg. Usually, one or two days later, dynamic exercises are added without weightbearing. Beginning on the eleventh day, the patient receives treatment in the swimming pool.

At the sixth week, stabilizing exercises for the toes, foot, and knee joints are begun with the patient sitting on the edge of the bed with elastic stockings on the legs. During breaks in exercise, the feet rest on an elevated board. All weightbearing during this phase must be avoided.

After the eighth week, provided there has been adequate training in circulation and endurance, the patient is allowed to move around in a wheelchair or with minimal weightbearing (contact with the ground) and to walk with crutches. Subsequent treatment follows standard nonsurgical guidelines.

Sample Exercises

The following exercises may be used in the early to middle phase of mobilization and strengthening for the injured leg. The patient exercises in a side-lying position on the normal side, with the normal leg in slight flexion. The therapist stands behind the patient, holding the injured leg, one hand under the foot, the other hand above the knee under the thigh:

- flexion and extension of the knee with the hip joint in a neutral position
- combined flexion of knee and hip joint, with extension
- attempting to hold the leg without help with the hip and knee in a neutral position
- abducting the extended leg
- bringing the extended leg slightly forward into hip flexion and back into extension or hyperextension (avoiding a lordosis)

Subsequently, as intensified exercises, the previous exercises are repeated, but without support, preferably exercising initially with a pillow between the legs rather than from an adducted position.

Throughout all these exercises, it is important to maintain a correct side-lying position, that is, a slight flexion of the lower leg with the upper hip in a neutral position, and not to wiggle. Resting in between exercises may also be useful.

Trochanteric and Subtrochanteric Fractures

The treatment of trochanteric and subtrochanteric fractures of the femoral neck are usually intensified more quickly. Frequently, these fractures tolerate weight-bearing two to four weeks earlier than fractures of the proximal and distal femoral neck.

TOTAL ENDOPROSTHESIS OF THE HIP JOINT

General Considerations

This section describes physical therapy for patients with endoprosthetic replacements in which plastic, metal, and bone cement have been used. We must distinguish between treatment using surgical techniques that require the removal of the greater trochanter and its refixation during the implantation from techniques that do not require this surgical manipulation.

In a "primary" total endoprosthesis, which is our main focus here, there is a synthetic replacement after a trauma to the hip joint, such as a fracture of the medial femoral head. A "secondary" total endoprosthesis is a treatment of choice in a severe arthrosis of the hip. The therapy for a secondary endoprosthesis differs from that for a primary replacement in only one aspect: it must deal with the fact that usually, prior to the operation, a considerable number of contractures (especially of the hip flexors, adductors, and the external rotators) have occurred and a compensatory lordosis and disturbed gait are apparent. The therapy must be provided for these preoperative conditions. In this context, therapy must be carefuly oriented to the following therapeutic limitations and restrictions.

Replacement with Dissection of the Greater Trochanter

The patient is placed with the bandaged leg in a flat abducted position, at first with a Charnley pillow, later with a wedge-shaped abduction pillow. The symptoms parallel those of a surgically treated fracture of the femoral neck. Similarly, the therapeutic treatment is the same as that for conservatively treated fractures of the proximal and distal femoral neck, except that there is no massage.

Course of Treatment

Preoperatively, the patient exercises as necessary following the guidelines for prophylactic treatment of bedridden patients as described previously, except that respiratory therapy is emphasized. The patient may also do isometric exercises with the injured leg. On the first postoperative day, the physical therapist applies the previously described prophylactic measures and isometric exercises for the injured leg. On the second day, the patient can be instructed in flexion and extension of the knee and hip joint while maintaining hip abduction. The therapist supports the leg to avoid unwanted stress to the surgical site.

Beginning on the fourth day, provided there are no complications in wound healing, the patient is allowed to sit on the edge of the bed and to begin standing up and using a walker. The legs must be protected by elastic stockings or elastic bandages.

Beginning on the eleventh day, additional mobilization and strengthening can be done in the swimming pool. The patient should be in the supine and prone positions and sitting in the water. Walking can be started at a high water level. Any rotation, abduction, and adduction must be avoided while exercising in and outside the water.

On the twelfth to fourteenth days, after successfully using the walker, the patient can switch to crutches. The gait is adjusted to the individual situation.

From the third week on, in addition to exercises in the supine position, and while sitting and standing (with the injured leg nonweightbearing), exercises in the prone position can be added to the program. The therapist has to support the patient's turning into a prone position in order to maintain the slight abduction of the injured leg.

Beginning in the fifth week, work in the various positions is intensified, and the crawling position is added. Stair climbing is also carefully begun. Beginning in the seventh week, exercise of the abductors is allowed. At first, the weight of the leg might have to be supported. The crutches are replaced by canes. Beginning in the eighth week, the transition from walking with a cane to normal walking on stairs is made. Beginning in the ninth week the leg can be adducted to the midline. While in the pool, the patient exercises the leg without restrictions.

Beginning in the fourth month, if the Trendelenburg sign is negative, a cane is no longer necessary. Beginning in the seventh month, the leg can be adducted across the midline. The patient is allowed to cross the legs when supine and while sitting.

Sample Exercises

The following exercises are used in the early phase of mobilization and strengthening. In the supine position:

- combined hip and knee flexion
- combined hip and knee flexion with head flexed or head and arms flexed
- flexing the legs: first knee toward the forehead and then flexing the leg, head, and arms; alternating the normal and injured side
- sitting up, depending on the patient's ability, using an assistive device or the patient's own muscle power
- flexing the normal leg to the abdomen, stretching the injured leg into extension (elevating the head to correct a possible lordosis)
- flexing the normal leg and extending the injured leg while elevating the bottom slightly from the bed with pressure from the normal leg.

All of the above exercises should be done in such a way that when tension is required it is built up and maintained from the foot.

The following exercises can be done while sitting on the edge of the bed:

- extension and flexion of the knee
- flexion of the hip and knee

The following exercises can be done while standing on the normal leg (using parallel bars or a rail):

- flexion of the hip and knee and then extension to the neutral position
- shifting the weight to the injured leg, maintaining extension during resistance at one or more points either separately or simultaneously

The above two sets of exercises may also be done in the swimming pool while supported by the buoyancy of the water.

Replacement without Dissection of the Greater Trochanter

The treatment is essentially the same as described above. However, abduction and adduction to midline are allowed at the beginning of the treatment, since the trochanter does not have to be temporarily removed and refixed again. The details of the exercise program should be coordinated with the physician in charge. Regardless of the surgical technique selected and whether it is a primary or secondary type of treatment, it is especially important not to overstress the patient. Instruction has to be adjusted and increased on an individual basis. Active exercises are most important. Passive manipulations should never be applied beyond the pain threshold of the patient. Electrotherapy, local massage, and the application of heat are contraindicated.

FEMORAL FRACTURES

Mechanism of Injury

Fractures of the femoral shaft are caused mainly by a direct impact. Depending on the type of force involved, the type of fracture will vary, but they usually involve bending, torsion, and compression.

Types

The following fracture types may be distinguished: torsion fractures, compound fractures, horizontal fractures, comminuted fractures, and comminuted fracture with several fragments.

Dislocations with shortening and considerable lateral deviation and separation of the bones are due to the pull of the strong thigh muscles, as observed, for instance, in femoral fractures (see Figure 7–3). Radiograms in two planes reveal the type and extent of the fracture.

Medical Management

Femoral fractures are accompanied by considerable blood loss (1 in 21 involve hemorrhagic shock) and require intensive care. The treatment of femoral fractures in children is usually nonsurgical, using a cast with traction (see Figure 7–4). In the first few years of age, the cast is just adhesive tape to the extended leg, usually with a 90 degree juxtaposition of hip and knee joint and with a slight shortening of the fragments to prevent excessive growth.

Nonsurgical treatment of a femoral fracture in adults requires an immobilization period in traction for several months and thereafter in a cast. Persistent problems

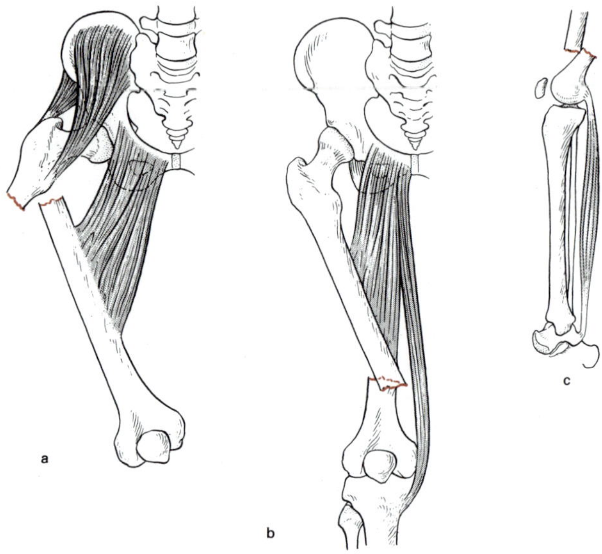

Figure 7–3 Displaced femoral fractures by muscle pull: a. displaced proximal femoral fracture; b. displaced distal femoral fracture in an AP-radiogram; c. in a lateral radiograph

Source: Schlosser and Kuner, 1980

with the range of motion in knee and hip joints and with malalignment and shortening have to be expected. An osteosynthesis, however, provides postoperative exercise stability and using a Kuntscher nail also provides early stability for weightbearing. Osteosynthesis also enables one to avoid malalignment and pseudarthroses.

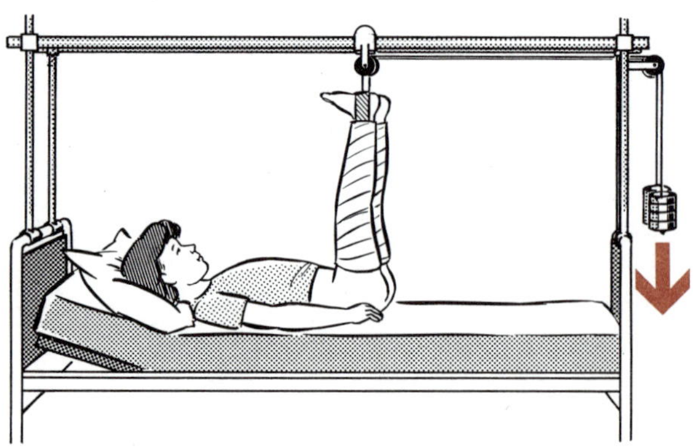

Figure 7–4 Vertical extension of both legs by traction with adhesive tape to compensate for shortening and a rotary malalignment due to femoral shaft fractures

The choice of osteosynthetic material depends on the type of fracture. Horizontal or short oblique fractures of the middle third of the femur indicate the need for an intramedullary nail. The intramedullary nail procedure is best done in a closed fracture. Fractures of the upper and lower thirds of the femoral shaft are usually fixed by a plate osteosynthesis. The plates are applied laterally, and medially there must be bony contact.

The stabilized fractures of the femoral shaft are treated initially by an active exercise program. The closer the fracture is to the knee joint the more this joint has to be flexed. Weightbearing has to be avoided for 8 to 16 weeks. In case of an intramedullary nailing, the nonweightbearing period should be 2 to 6 weeks.

Acute and late complications in nonsurgical treatment, malalignment and pseudarthroses are possible, and an infection might occur after an osteosynthesis.

Physical Therapy

The following therapeutic program is applicable to a femoral fracture stabilized by an intramedullary nail.

Symptoms

With the bandaged leg placed in a splint, the following symptoms are observable:

- usually a minor postoperative hematoma
- small scar in the area of the greater trochanter
- general activity slightly reduced
- pain around the fracture site

Therapeutic Aspects

- pneumonia and thrombosis prophylaxis: (1) respiratory therapy, (2) isometric and dynamic exercises
- circulatory stimulation: (1) isometric and dynamic exercises, (2) changing positions, (3) therapeutic pool
- prevention of edema: (1) elevation, (2) compression, (3) isometric and dynamic exercises
- mobilization: (1) isometric and dynamic exercises, (2) relaxation techniques as necessary (ice packs), (3) therapeutic pool
- strengthening: (1) dynamic exercises, (2) exercises against resistance, with or without equipment, (3) therapeutic pool, (4) recreation therapy
- activities of daily living: (1) walking exercises, (2) training in activities of daily living, as applicable

Course of Treatment

After the first postoperative day, the patient exercises using the guidelines for prophylactic treatment of bedridden patients, as described earlier. The patient also begins isometric muscle contractions with the injured leg.

After the fourth day, dynamic exercises in supine, side, and prone positions are added. In this early phase, the treatment must avoid any rotation.

Usually, healing of the wound is completed after the tenth postoperative day. At that point, the patient is allowed to swim in the swimming pool. With a high water level, the patient practices standing and walking and is also allowed to bear weight outside the pool, with the objective of progressing as soon as possible to a four-point gait pattern with crutches. Loading to the femur is allowed only through its longitudinal axis such as in standing and walking. During the exercises, manual resistance is applied below the fracture only after complete consolidation of the fracture.

Sample Exercises

The following exercises are used in the middle phase of healing to strengthen the gluteal muscles while the patient is in a prone position with the legs hanging down from the edge of the supporting surface:

- elevating the legs in extension slowly from the floor up to the upper-body level and back down
- moving the injured leg slowly up and down
- moving both legs alternately up and down
- elevating both legs in extension, spreading them and moving them back down in this position
- elevating both legs in extension, spreading them wide and closing them, then moving them back down
- spreading the legs wide on the floor, elevating them in a spread position, closing them, then moving them back down

The following exercises are used to strengthen the quadriceps muscle, with the patient sitting at the edge of the supporting surface:

- moving the injured leg in knee extension with the foot in a neutral position, then to doriflexion and inversion, then to dorsiflexion and eversion, then to plantarflexion
- moving both legs alternately in knee extension and flexion (moving the feet as indicated above)

- moving both legs at the same time in knee extension and flexion (moving the feet as indicated above)
- with a foam pillow or ball between the feet (distal tension), elevating in knee extension and back

The following exercises are used to strengthen the entire leg in a supine position:

- elevating and lowering the leg (moving the foot as indicated above)
- slightly flexing the hip, extending the knee and moving the leg back down in extension
- bringing both legs (pillow or ball between the feet) into flexion and back into extension (to prevent lordosis: elevating the head or, if necessary, supporting it with forearms)
- bringing both legs (with pillow or ball between the feet) into flexion, extending the knees and then lowering the legs with correction for lordosis as in previous exercise)
- elevating both legs, with pillow or ball between the feet, in an extended position, and then moving them back down (with correction for lordosis as previous exercise)

Physical Therapy

The following therapeutic program is applicable to a fracture of the femoral shaft stabilized by plating.

Symptoms

The injured leg is bandaged and placed in a Kirschner splint. The splint is set for two hours in 90 degrees of hip and knee flexion and then for two hours with the best possible extension (see Figure 7–5). After three days, the splint remains in the neutral position; usually, it is removed after the tenth day. The following symptoms are observable:

- postoperative hematoma, sometimes local elevated temperature
- long lateral scar with an extreme tendency to adhesions
- doughy turgor
- possibly decreased blood flow in the leg and in areas of the body against the bed
- possibly decreased general mobility and accordingly decreased respiratory and circulatory functions
- pain around the surgical incision

Therapeutic Aspects

- pressure area prophylaxis: (1) isometric and dynamic exercises, (2) changing positions
- pneumonia and thrombosis prophylaxis: (1) respiratory therapy, (2) isometric and dynamic exercises
- circulatory stimulation: (1) active exercises, (2) changing positions, (3) therapeutic pool
- prevention of edema: (1) elevation and compression, (2) isometric and dynamic exercises, (3) possibly ice application
- mobilization: (1) active exercises, (2) possibly use of an assistive dynamic splint (Frankfort design), (3) relaxation techniques, (4) cryotherapy, (5) possibly connective tissue massage
- strengthening: (1) dynamic exercises, (2) exercises against resistance, with or without equipment, (3) recreational therapy, (4) therapeutic pool
- training in activities of daily living: (1) walking exercises, (2) possibly training in other activities of daily living

Course of Treatment

In the first three days, the patient exercises following the guidelines for prophylactic therapy of bedridden patients, as described earlier. The patient does isometric muscle contractions of the injured leg with the splint in extension.

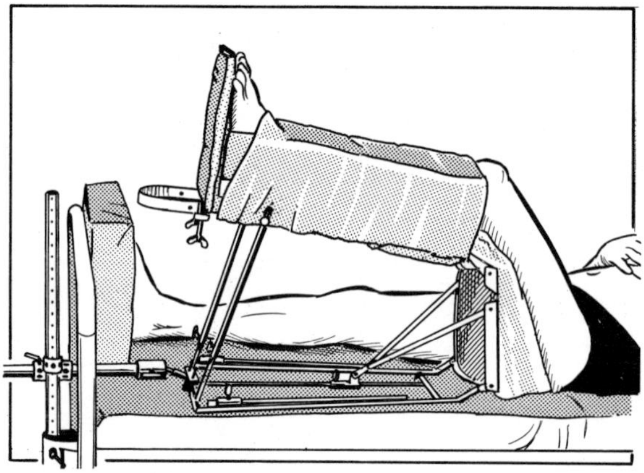

Figure 7–5 Hip and knee flexion at 90 degrees

Beginning on the fourth day, the splint is adjusted to a neutral position. Dynamic exercises are allowed, with the patient in a supine position and the therapist initially supporting to avoid stress to the fracture site. On the eleventh day, the splint is removed. The patient now exercises in side and prone positions (especially for exercising the gluteal muscles) and while sitting on the edge of the bed. The patient also participates in hydrotherapeutic exercises. In case of persistent problems with flexion of the leg, additional work in a dynamic splint is useful.

After about the third postoperative week, the patient walks without, or with only minimal, weightbearing using crutches (and elastic stockings). In the twelfth week, walking is begun in the pool. During the thirteenth week, weightbearing (10–20 kg) is started outside the pool with the rate of increased weightbearing and the withdrawal of walking aids adjusted to the individual situation.

Sample Exercises

The following exercises are used in the early phase of circulatory stimulation, mobilization, and strengthening, with the splint in extension:

- actively moving the ankles and toes
- isometric exercises of the quadriceps muscle and ankle: (1) dorsiflexion and slight contraction of the rectus femoris of the quadriceps, (2) plantarflexion/inversion and slight contraction of medial head of the quadriceps, and (3) plantarflexion/eversion and slight contraction of the lateral head of the quadriceps
- in approximately 30 degrees of knee flexion (adjusting the splint by hand or foam pillow underneath the knee), dynamic contractions of the quadriceps muscle, including the foot as in the exercise above, and with the lower leg partially supported initially by the therapist
- elevating the leg in extension, including the foot as in the above exercise, with partial support initially

The following exercises are done without a splint, with the patient in a supine position:

- all of the exercises described previously
- if there is good control of the previous exercises and no restrictions imposed by the physician, then exercising in diagonal movements of PNF can be done (prior to complete consolidation and without the use of resistance, traction, pressure, or other stressful manipulations)

DISTAL FEMORAL FRACTURES

Mechanism of Injury

Femoral fractures close to the knee joint are caused by a direct force, which may result in a combination of bending and rotational forces to the femur.

Types

Supracondylar and pericondylar femoral fractures are two types of distal femoral fractures close to the knee joint (see Figure 7–6). The latter type is considered to be a joint fracture. Comminuted fractures with several or multiple fragments occur mainly in bicycle accidents. In a supracondylar femoral fracture that does not involve the knee joint, the distal fragment is pulled by the muscles of the lower leg into a recurvatum. The tilted distal fragment can compress and damage nerves and blood vessels.

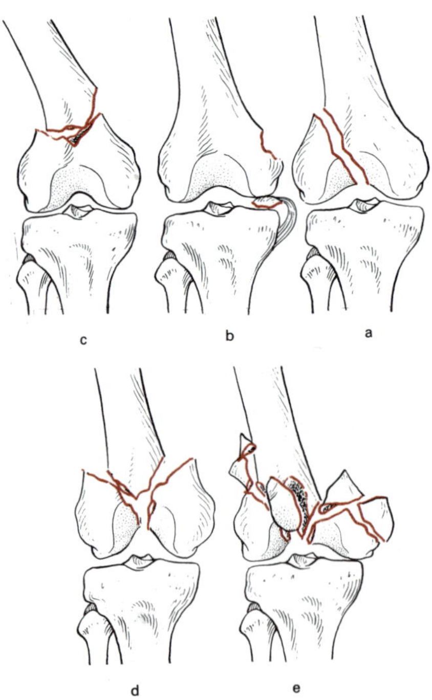

Figure 7–6 Types of distal femur fracture: a. longitudinal fissure; b. segment of bone is torn away; c. transverse fracture above the condyles; d. y-fracture; e. multiple-fragment fracture

Condylar and pericondylar fractures often result in multiple T- or Y-shaped fragments, with fracture lines extending into the intra-articular surfaces of the knee joint.

Clinical Considerations

The symptoms of a distal femoral fracture are local swelling and, in pericondylar fractures, an intra-articular hematoma, abnormal mobility, crepitation, pain interfering with function, axial deviation, and possibly neurologic and vascular disturbances.

Diagnosis

Radiograms in two planes reveal the type and the severity of the fracture.

Medical Management

A supracondylar fracture is usually treated surgically; a pericondylar fracture is always treated surgically. A minor dislocation of bone fragments can be stabilized sufficiently by screws. Dislocated fractures require reduction and fixation with condylar plates, with additional screws required in fractures with several or multiple bony fragments. Remaining defects are filled with autologous spongy bone. Minor contusion fractures can also be treated with special dynamic splints and, in adolescents, with a cast. If in a child's fracture the epiphyseal plate is involved, the fracture has to be reduced and stabilized with screws.

Acute and Late Complications

Injuries to blood vessels and nerves are early complications that must be treated first. Arthroses due to incorrectly reduced joint fractures and malalignment with an incongrous arthrosis may appear as later complications.

Physical Therapy

Basically, surgically treated supracondylar fractures (mostly stabilized by condylar plates) are treated in the same way as femoral shaft fractures that are stabilized by plates. However, in supracondylar fractures close to the joint, the mobilization of the knee and the training of the extensor muscles is more difficult, and there is a high risk of an effusion caused by irritation.

Chapter 8

Injuries of the Knee Joint

The knee joint is a complicated joint whose anatomical structure characteristics, combined with considerable static and functional load, account for the high frequency, variety, and severity of injuries. Relatively early, especially after a trauma, stresses on the knee joint cause degenerative changes and wear on the joint cartilage and menisci.

Distal femoral fractures, and also fractures of the tibial head and patella, affect the knee joint. They often occur with lesions of the internal structures and ligaments. Depending on the type of force involved (direct or indirect), injuries of the soft tissue and the bony components can occur separately or in combination (see Figure 8–1).

PATELLA FRACTURE

Mechanism of Injury

The patella is a sesamoid bone located between the tendon of the quadriceps muscle and the patellar ligament. During the flexion of the knee joint, it glides over the femoral condyles (femoropatellar joint). A fracture of the patella is caused by a fall on the knee or by an impact on the patella by a direct force (for example, during a car accident). Excessive and uncoordinated tension of the extensor muscles of the thigh together with flexion of the knee joint can result in a tearing fracture by internal forces. A bone fragment can also be endogenously torn out of the tibial tuberosity. Ruptures of the patellar ligament and the tendon of the quadriceps muscle are caused mainly by preexisting degenerative lesions.

Types

Undisplaced horizontal fractures occur with the extensor apparatus intact, but usually the aponeurosis of the extensor tendon is ruptured, which contributes to a

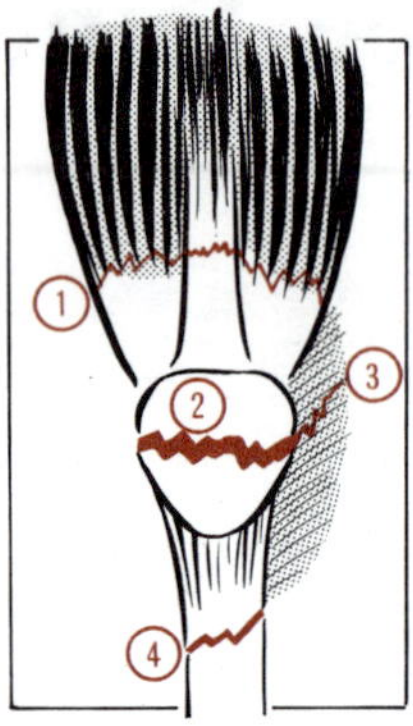

Figure 8–1 Injuries of the extensor muscles of the knee joint: a. ruptured tendon of the quadriceps muscles; b. patellar fracture; c. ruptured retinaculum; d. ruptured patellar tendon

dislocation of the bony fragments. Star-shaped and comminuted fractures are observed after a direct impact.

Clinical Considerations and Diagnosis

The main symptom of patella fracture is the loss of the knee joint's ability to extend itself. The fracture of the patella is frequently combined with an effusion into the joint. The horizontal displaced fracture is characterized by a depression, which can be palpated. The diagnosis is confirmed radiographically. A congenital patella bi- or tripartita has to be excluded as a differential diagnosis, as it may mimic a fractured patella.

Medical Management

The fragments must be adapted carefully and fixed by wire cerclage and/or screws. Small fragments that cannot be put back in place should be removed. In a comminuted fracture, removal of the patella is usually not necessary. Undisplaced patellar fractures with intact aponeurosis of the extensor tendon, called sub-aponeurotic fractures of the patella, can also be treated nonsurgically by cast immobilization.

Acute and Late Complications

Fractures that are healed in a dislocated position can result in insufficient extensor muscle function. Also, uneven areas on the posterior surface of the joint can cause a femoropatellar arthrosis.

Physical Therapy

The following therapeutic program is applicable to nonsurgically treated fractures of the patella.

Symptoms

With the injured leg placed in a split cast or a splint, the following symptoms are observable:

- usually discoloration by a hematoma, swelling and effusion around the knee joint
- general inactivity and, therefore, reduced respiratory and circulatory function
- pain around the knee

Therapeutic Aspects

- pressure area prophylaxis: (1) changing positions, (2) isometric and dynamic exercises
- pneumonia and thrombosis prophylaxis: (1) respiratory therapy, (2) isometric and dynamic exercises
- circulatory stimulation: (1) respiratory therapy, (2) active movement, (3) changing positions, (4) therapeutic pool
- edema prophylaxis: (1) elevation and compression, (2) isometric and dynamic exercises
- mobilization: (1) active movement, (2) relaxation techniques with ice, (3) therapeutic pool, (4) changing positions, (5) possibly use of an assistive dynamic splint (Frankfurt design)
- strengthening: (1) dynamic exercises, (2) moving against resistance with or without equipment, (3) electrotherapy (i.e., intermittent galvanic current), (4) therapeutic pool, (5) recreation therapy
- training in activities of daily living: (1) walking activities, (2) possibly exercise in other skills of daily living

Course of Treatment

Beginning on the first day, the therapist motivates the patient to exercise intensively, following the previously described guidelines for prophylactic therapy for bedridden patients. On about the eleventh day, isometric contractions of the quadriceps muscle are begun. The leg remains in the splint cast. In the third week, dynamic exercises can be started. Strengthening of the quadriceps muscle through full extension is important. Full knee flexion, preferably accompanied by ice applications to prevent edema, requires a great deal of time. Careful exercises, especially knee flexion, are recommended because of the high risk of effusion.

After the sixth week, the patient is allowed to go into the therapeutic pool. If the radiographic results are satisfactory, weightbearing is usually started in the tenth week. Initially, the knee is protected by a knee orthosis. The patient walks with crutches weightbearing to 10 kg. The weight is increased in the following weeks, depending on the individual situation. Finally, swimming, occupational therapy, recreation therapy, and training in a correct gait and posture can be added to the treatment schedule.

Sample Exercises

The following exercises can be used in the middle phase of mobilization and strengthening. Exercises in the therapeutic pool:

1. sitting in the water
   - extending the injured leg, ensuring that the thigh does not float to the surface, then flexing the knee
   - the same exercise, with both legs symmetrically
   - the same exercise, with legs alternately
2. while standing on the normal leg, with injured leg in 90 degrees of hip and knee flexion and one hand underneath the thigh at knee level: knee extension and flexion including the ankle, alternating medial dorsal extension dorsiflexion, plantarflexion, eversion, inversion
3. supine position
   - flexion and extension of the hip and knee unilaterally, bilaterally and symmetrically, bilaterally and alternately
   - adducting and abducting the extended leg, unilaterally and bilaterally
   - elevating the extended leg slightly out of the water, unilaterally and bilaterally
4. prone position, holding on to the edge of the pool
   - both hips flexed, isolated knee flexion and knee extension
   - combined hip knee flexion and extension

- flexing both knees, pushing feet downward, with extended legs floating up to the surface
- moving the extended legs alternately, up and down

5. standing

- knee flexion and extension
- hip and knee flexion, then knee extension, then pulling it back to the normal leg in extension
- the same exercise, but extending to the side
- the same exercise, but hyperextending the hip

Patellar Fractures Fixated by Wire Cerclage

Symptoms

In a patellar fracture fixated by wire cerclage, the symptoms are the same as in a nonsurgically treated fracture of the patella, except that (1) frequently, the leg is not placed in a splint cast but only bandaged in a splint, and (2) medially to the patella there is a longitudinal scar, which often adheres to the underlying tissue.

Therapeutic Aspects

The therapeutic aspects are the same as in a nonsurgically treated fracture of the patella. However, after the wound is completely healed, a special treatment for the scar (ointments, massage) may be added to the above exercises aimed at supporting blood circulation and mobilization.

Course of Treatment

In the first few days after the operation, the patient exercises following the previously described guidelines for prophylactic therapy for bedridden patients. The patient also learns to do isometric contractions of the quadriceps muscle of the injured leg. After about the fourth day, the treatment can be expanded to include dynamic exercises. Especially during flexion exercises, the risk of irritation is increased, which requires precaution and preventive ice treatment.

At the third week, mobilization and strengthening exercises are begun, paralleling the procedures outside the pool. After the tenth week, the treatment is the same as in nonsurgical therapy.

Sample Exercises

The following exercises are used in the middle phase of mobilization and strengthening. Flexion exercises in the therapeutic pool:

1. prone position, holding on to the edge of the pool

    - hip flexion, feet moving toward the pool bottom, unilaterally, bilaterally, alternately
    - combined flexion of knee and hip, knee extension downward, bilaterally and symmetrically
    - the same exercise, and then extension to the surface, bilaterally and symmetrically alternating the preceding two exercises

2. Standing on the normal leg, facing the edge of the pool

    - big toe moving up the wall until the best possible knee flexion is reached

3. standing, with the back toward the edge of the pool

    - hip and knee flexion and extension
    - hip extension-abduction-internal rotation to flexion-adduction-external rotation
    - grasping the edge of the pool, flexing both legs to waist height

4. supine position: the same preceding three exercises
5. sitting on the edge

    - with thighs remaining on the edge: knee flexion and extension, unilaterally, bilaterally and symmetrically, then alternately
    - "bicycling" forward and backward
    - knee flexion against a buoy

Removal of the Patella

Symptoms

The symptoms in a removal of the patella are very similar to those in a nonsurgically treated fracture of the patella. Usually, there is a scar across the knee, and a great many adhesions can be seen in the surrounding tissue.

Therapeutic Aspects

The therapeutic aspects parallel those for a patellar fracture stabilized by wire cerclage. Here, however, the treatment is concerned mainly with strengthening the knee for stability, rather than mobilizing it.

Course of Treatment

After the first postoperative day, the therapist works with the patient, following the previously described guidelines for prophylactic treatment of bedridden patients. Beginning on the eleventh day, intensive isometric contractions of all portions of the quadriceps muscle are added to the treatment. Beginning in the sixth week, dynamic exercises are allowed. Initially, extension exercises are more important than flexion exercises. In addition to physical therapy, the use of intermittent galvanic current can facilitate this treatment program. Careful mobilization and application of ice can help to prevent an effusion into the joint.

After the seventh week, the patient begins to exercise in the therapeutic pool. In the thirteenth week, weightbearing begins. Henceforth, the program is the same as for a nonsurgical treatment.

Sample Exercises

In the early phase of the circulatory support, mobilization, and strengthening, the therapist prevents atrophy with a resistive tubing. Initially, this is done under the therapist's supervision; later, the patient does it alone. Resistive exercises with the tubing are done ten times in succession, four to six times a day. In a supine position, tubing fixed to an overhead frame, the patient grasps the tubing and moves the arm from shoulder flexion and adduction with elbow flexion and supination to extension, abduction, and pronation. The patient then grasps the tubing and moves the arm from a slight shoulder flexion-adduction-external rotation with elbow extended to an extension-abduction-internal rotation with elbow extended. Both of these exercises, when done with dumbbells, tubing, and pulleys are well-suited to strengthening the arms, especially in preparation for walking with crutches.

The following exercises are done with the patient in a supine position, injured leg in a splint, the normal leg being exercised:

- To exercise from mid-hip/knee flexion to an extension (depending on the patient's strength at this point), the patient places the tubing under the sole and grasps the tubing, then extends the leg against the resistance of the tubing without moving the hands, maintains the extended position a short while, then flexes again.
- The same exercise (to the extent possible with the splint) from mid-hip/knee flexion to extension-adduction-external rotation with the foot in dorsiflexion, then to extension-abduction-internal rotation with the foot in plantarflexion.
- Tubing fixed to a frame above the knee, the knee is flexed approximately 50 degrees, hanging in the loop of the tubing. The leg is now extended with

the foot in dorsiflexion while maintaining the extension, then returning to the initial position.

- The tubing is fixed to a frame above the foot of the bed; the heel rests in the tube loop, with the leg suspended in about 40 degrees of hip flexion. The heel is then brought down, with the foot dorsiflexed and the knee extended, maintained in extension, then returned to the initial position.
- The same exercise (to the extent possible with the splint), with the foot in dorsiflexion, moving the leg to extension-adduction-external rotation and with the foot in dorsal plantarflexion, moving the leg to extension-abduction-internal rotation.

In doing the last three exercises, one should adjust the tension to avoid genu recurvatum and be particularly careful in the case of patients with lumbar problems.

INJURIES OF THE INTERNAL STRUCTURES OF THE KNEE JOINT

As noted earlier, the knee joint is especially at risk because of its structure, function, and thin skin cover. The knee is similar to a hinge joint, moving in a complicated rolling-sliding motion. The movement is performed by a complex ligament muscle system without any mechanical protection provided by the shape of the bony joint components.

Mechanism of Injury

An indirect trauma is caused by a force that exceeds the physiological range of motion and resistant capacity of the soft tissue. The cartilage, bone, and the relatively unprotected joint capsule with its mucosal lining are also frequently affected by a direct impact.

Types

Mechanical forces that hit the ligaments and other internal structures of the knee joints affect not just one structure but several in sequence. Thus complex ligament ruptures are more common than solitary ones. There are a variety of such injuries, ranging from simple contusion or rotation trauma, with or without ligament rupture or lesions of internal structures, to injuries of bone, ligament, and cartilage. Also, the knee joint has a tendency to react to irritations to the joint structures resulting in lesions. The lesions can be caused by preexisting degenerative changes. Such changes affect the knee joint more often than any other body joint.

Contusion

Mechanism of Injury

The contusion is a common injury, caused by a blunt, direct force on the knee. As a result, the soft tissue is squeezed against the bony structures.

Types

Circumscribed or extensive hematomas are caused by ruptured blood vessels. Hematomas between the fasciae when combined with tissue necrosis can lead to an ossifying myositis.

Hemorrhages also affect the synovial membrane, especially in the area of the capsular folds. The extent of the hemorrhage is decisive in determining the course of treatment. The symptoms of contusion can subside within a few days, but an extensive hemorrhage can cause secondary irritation, especially in connection with preexisting degenerative changes.

Medical Management

An acute contusion is treated by immobilization, ice packs or spray, and a compressive bandage. Additional medication is given to stop the swelling and improve fluid resorption. Afterward, a reactive hyperemia and isometric muscle exercises are initiated. In addition to the joint injury, the possibility of nerve lesions must be considered; such lesions can be caused either by the impact or by indirect compression of a nerve branch.

Rotation Injuries

Rotation injuries vary in severity, depending on the strength and type of force. When rotation exceeds the normal range of motion, the result is capsule and ligament lesions.

Types

The clinician distinguishes between a mild and a severe rotation injury. A severe rotation injury can be accompanied by considerably more lesions of the ligaments than in a subluxation. In a mild rotation injury, however, the ligaments remain intact. Such an injury, also called overstretching, can appear in combination with ruptured capsular blood vessels.

In a moderate rotation injury, the ligaments are affected to a greater degree than in a mild injury, but the continuity of the joint is still preserved. Positioned radiograms can confirm a diagnosis of intact ligaments. General symptoms of the injury are swelling, pain, and a limited range of motion.

Clinical Considerations

The symptoms of a rotation injury are similar to those of a mild injury, but they can be more distinctive.

Medical Management

The treatment of a mild injury parallels that for a contusion. A moderate injury might require temporary immobilization, followed by carefully designed exercise therapy.

Physical Therapy

The following therapeutic program is applicable to a rotation injury of the knee joint (without clinical lesions of the ligaments and the meniscus).

Symptoms

With a compression bandage (sometimes a cast) around the knee and the knee placed in a splint, the following symptoms are observable:

- swelling, possibly effusion
- decreased general activity
- possibly also decreased respiratory and circulatory functions
- pain around the knee joint

Therapeutic Aspects

- pressure area prophylaxis: (1) isometric and dynamic exercises, (2) changing positions
- pneumonia and thrombosis prophylaxis: (1) respiratory therapy, (2) isometric and dynamic exercises
- circulatory stimulation: (1) respiratory therapy, (2) active movement, (3) changing positions, (4) cryotherapy, (5) therapeutic bath
- edema prophylaxis: (1) elevation, (2) compression, (3) isometric and dynamic exercises
- mobilization: (1) active movement, (2) relaxation techniques with ice, (3) therapeutic bath, (4) changing positions
- strengthening: (1) dynamic exercises, (2) moving against resistance, with or without equipment, (3) hydrotherapy, (4) recreational therapy
- training in daily routine: gait exercise

Course of Treatment

On the first day after the injury, the therapist begins to work with the patient, following the previously described guidelines for prophylactic treatment of bed-ridden patients. Immobilization of the injured leg must be maintained.

By the third day at the latest, the therapist removes the compression bandage and has the patient begin contractions of all parts of the quadriceps muscle; this is combined with ice therapy. After this treatment, the compression bandage is put back in place and the leg is again placed in the splint. After the effusion subsides, the remaining effusion or possible capsular swelling may be diminished further by various modalities.

The strengthening of the muscles, especially the quadriceps muscle, is the next objective of the therapy program. The application of ice helps to prevent a recurrent effusion. Flexion exercises must be done carefully to avoid irritation and overstretching on the already affected capsule.

If there are no further complications and there is sufficient muscular support, the patient is allowed to get up and start weightbearing. This should be done initially with an orthotic support and crutches using a three-point gait pattern. If the knee remains free of symptoms, these supporting devices are gradually withdrawn.

The therapist continues to work with the patient on mobilization and strengthening. These activities can be supplemented by exercises in the therapeutic pool, later by swimming. Finally, exercises against controlled resistance, electrotherapy, occupational therapy, and recreational therapy are added to the therapeutic program.

Sample Exercises

The following exercises are used in the late phase of strengthening:

1. with resistive tubing
   - alternate flexion and extension of knee and hip
   - symmetrical flexion and extension of knee and hip
   - extended legs, alternating up and down
   - abducting and adducting the extended legs
2. using an exercise bicycle, at first with elevated seat and little resistance, then increasing the resistance
3. PNF
   - prone position: pattern for knee extension and flexion
   - supine position: both diagonals, with the knee flexed and extended, starting carefully with resistance, stretching, pulling, and pressure

Rotation with Ruptured Ligament

Mechanism of Injury

Ligament ruptures are caused by indirect forces.

Types

Severe rotational injuries are accompanied by partial or complete ruptured ligaments.

Diagnosis

The extent of the ligament lesion is verifiable clinically and radiographically. The severity of the lesion is indicated by lateral dislocation (spreading of the joint space with a lesion of the lateral ligaments, by anterior or posterior drawer signs with ruptured cruciate ligaments, and in complex injuries by rotary instability). Usually, there is a hematoma, which if the posterior capsule is ruptured, might drain into the soft tissue of the lower leg. In such cases, the wound's appearance might mask the severity of the injury and result in inadequate treatment.

The following types of ruptures are distinguishable in a single plane:

- rupture of the medial ligament with valgus position
- rupture of the lateral ligament with varus position
- rupture of the anterior cruciate ligament with anterior shifting of the lower leg
- rupture of the posterior cruciate ligament with posterior shifting of the lower leg

Usually, the ruptures are in two planes. These ruptures involve:

- anteromedial instability with a lesion of the medial collateral ligament, anterior cruciate ligament, and medial posterior capsule
- anterolateral instability with a lesion of the lateral collateral ligament, anterior cruciate ligament, and lateral posterior capsule
- posteromedial instability with a lesion of the lateral collateral ligament, posterior cruciate ligament, and posterior medial capsule
- posterolateral instability with a lesion of the lateral collateral ligament, posterior cruciate ligament, and posterior lateral capsule

A rotatory injury of the knee joint might involve other internal structures, especially the menisci. This has to be considered in diagnosis and treatment.

Medical Management

Ruptured ligaments and complex injuries require immediate surgical treatment to restore the stability and function of the joint. A direct suture or refixation of a displaced bone fragment is indicated, depending on the site of the injury. Postoperatively, complete immobilization for six weeks or incomplete immobilization in a walking cast for three to six weeks is necessary.

Ligament injuries require careful but intensive physical therapy, regardless of whether they have been treated nonsurgically, with a lengthy period of immobilization, or surgically. In case of a pain and/or effusion, the therapy must be supplemented with ice packs and compression bandages, when the cast is off. A persistently blocked joint might have to be mobilized under general anesthesia. The nature and extent of the injury determines the course of treatment.

Late Complications

Incorrect treatment of a fresh ligament injury frequently leads to ligamentous instability that later requires plastic surgery. Such surgery may in turn require lengthy and intensive physical therapy.

Old Ligament Injuries

Ligament injuries that have been treated incorrectly, incompletely, or inappropriately will remain unstable.

Types

Chronically unstable ligaments have the same symptoms as early lesions of the ligaments.

Clinical Considerations and Diagnosis

Examination of the function of the collateral ligaments in extension and slight flexion, of the drawer sign in sagittal and rotary directions, and symptoms of meniscus lesions will reveal the type of chronic instability. Frequently, the examination also reveals chronic irritation of other joint structures.

Medical Management

The surgical procedures required to reconstruct persistent instability of the knee joint are aimed at repairing the stabilizing structures (cruciate and collateral ligaments, posterior capsule) and at tightening the ligaments by reinserting them at different points. Tendons and ligaments are also deviated and used both as active stabilizing structures and as supports for passive stability of the knee joint.

Given these basic components, the physical therapist must take into consideration two opposite therapeutic aspects: (1) the joint needs a period of immobilization to give the reinserted and sutured ligament time to reconnect to the surrounding tissue (this takes about six weeks in a bradytrophic fascial and tendinous tissue), and (2) exercises of the dynamic stabilizing components are necessary from the beginning.

After the wound pain has subsided sufficiently, intensive isometric exercises must be started, with the cast still in place. Isometric exercise by electrical stimulation of the muscles can enhance or replace active exercises. The quadriceps muscle is the most important; it stabilizes the joint by extension, not only anteriorly, but over its lateral portions merging with both retinacula laterally.

After plastic repair, the muscles have a tendency to become weak and functionally useless. Therefore, special isometric muscle training is a priority. After the removal of the cast four to six weeks postoperatively, these exercises are continued and then replaced by dynamic exercises. The full range of motion is not yet desired because it would interfere with the healing process of the ligament and endanger the new insertion points. The full range of motion should not be reached earlier than 6 to 9 months. Athletic activities can be started after 9 to 12 months.

Physical Therapy

The following therapeutic program is applicable to suture repair of the medial collateral ligament of the knee.

Symptoms

With the injured leg placed in a cast or a splint, the following symptoms are observable:

- postoperative edema, possibly also painful effusion from inactivity
- general inactivity
- decreased circulatory function and decreased blood flow due to pain and supine position
- pain

Therapeutic Aspects

- pressure area prophylaxis: (1) isometric and dynamic exercises, (2) changing positions
- pneumonia and thrombosis prophylaxis: (1) respiratory therapy, (2) isometric and dynamic exercises

- circulatory stimulation: (1) active movement, (2) changing positions, (3) cryotherapy, (4) therapeutic pool
- edema prophylaxis: (1) elevation, (2) compression, (3) isometric and dynamic exercises
- mobilization: (1) active movement, (2) relaxation techniques with ice, (3) therapeutic pool, (4) changing positions
- strengthening: (1) dynamic exercises, (2) exercises against resistance with or without equipment, (3) therapeutic pool, (4) recreation therapy
- activities of daily living: gait training

Course of Treatment

During the first ten postoperative days, the patient exercises following the previously described guidelines for prophylactic treatment of bedridden patients. If the healing is progressing without any complications, a cast is applied on the eleventh day, and the patient is sent home for four to five weeks.

At the beginning of the seventh postoperative week, at the earliest, the cast is removed and physical therapy is started. After edema prophylaxis, the knee joint is exercised to full extension, with special strengthening of the vastus medialis and obliques muscle.

To increase knee flexion, careful mobilizing exercises are tried in various positions, together with ice applications to prevent effusion. The patient also works on mobilization and strengthening in the therapeutic pool. In case of any irritation and effusion, the exercise program must be reduced or suspended.

Beginning in the eighth week at the earliest, weightbearing begins. During the first days of weightbearing, the patient walks at a high water level in the therapeutic pool. Outside the pool, the patient walks with an orthotic knee splint (and crutches using a three-point gait pattern).

Depending on the functional and muscular status, the weightbearing is increased to full weightbearing while walking. Stabilizing exercises should be done in a standing position. In addition, balancing and strengthening exercises on a balance board are recommended. Swimming, recreational therapy, and occupational therapy complete the rehabilitation program.

Sample Exercises

The following exercises can be used in the late phase of strengthening. The first set of exercises is performed with a foam block or ball between the feet:

- while in a supine or sitting position keeping the foam or ball between the feet, ''holding'' it against resistance

- while sitting on the edge, knee extension and flexion, "holding" the foam block or ball at different levels against resistance

The following set of PNF exercises are done in the supine position:

- from extension-abduction-internal rotation to flexion-adduction-external rotation, with the knee in flexion
- the same exercise diagonally, with the knee in extension

The following set of exercises is done with weights:

- sitting: knee extension and assisted flexion with various movements of the foot (up, up and inverted, up and everted)
- supine or semi-sitting position: flexion, extension, hip flexion with knee extension, then back down, then with hip abduction and adduction
- supine or semi-sitting position: all diagonal lower-extremity exercises previously listed

Suture Repair of the Cruciate Ligaments of the Knee

Symptoms, therapeutic procedures, and the course of treatment up to the seventh postoperative week are the same as for suture repair of the medial collateral ligament. After the removal of the cast and initial circulatory support (elevation, compression, isometric muscle contractions), careful mobilization of ankle and knee joint and strengthening of the muscles are begun.

During the first few days, the knee is mobilized primarily with an assistive dynamic splint (Frankfurt design, see Figure 8–2), accompanied by ice applications. Afterward, the treatment is continued using the usual manual techniques. At the same time, the muscle strength is built up. This is intensified as the mobilization continues. During the training of the quadriceps muscle (avoiding hyperextension), it is also important to strengthen the knee flexor muscles. Here it should be remembered that flexors of the knee include not only the hamstring muscles but also the gastrocnemius muscle. Subsequent swimming in the therapeutic pool supports the mobilization and strengthening exercises.

At the end of the rehabilitation period, with exercises using PNF techniques, stretch and approximation exercises should either be eliminated completely or, if used, applied with great care.

The nonweightbearing period for patients with a lesion of a cruciate ligament can last up to three months. The rate of increase in weightbearing will depend on the individual situation, for example, muscular condition, irritation, and so on (see also the discussion on suture repair of the lateral collateral ligament).

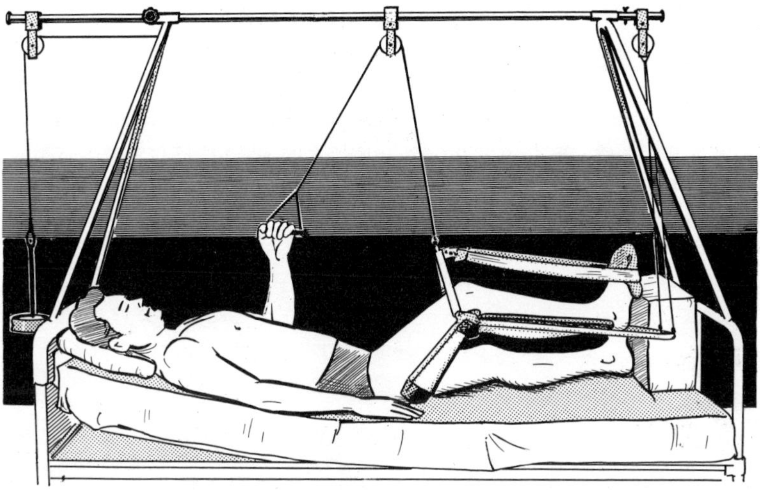

Figure 8–2 Frankfurt traction apparatus

For patients with plastic repair of the ligaments (especially after chronic lesions), the treatment is conducted with great care and applied more slowly. Weightbearing should not be started before three to four months.

Sample Exercises

The following exercises are used in the late phase of strengthening and gait training:

1. PNF in prone position: extension, and especially flexion pattern
2. PNF in supine position

   - both diagonals, each with an extended and flexed knee
   - both diagonals, each with an extended knee

3. Exercises in parallel bars

   - standing up, holding the position during resistance (especially to pelvis-knee and knee-toes)
   - standing up, alternately flexing and extending the knee (approximately 0–30 degrees), first exercising the injured leg with fixed normal leg, then moving the legs alternately

MULTIPLE SOFT-TISSUE INJURIES AROUND THE KNEE JOINT

Physical therapy for injuries of a single soft-tissue structure in the knee joint was discussed earlier in this chapter. The physical therapist who is able to apply that therapy will now be able to combine such individual therapeutic treatments in cases of multiple lesions. The deciding factor in designing the therapeutic program (onset of therapy, types, and increase of therapeutic exercises) will be determined by the most severe traumatic component.

SUBLUXATIONS

Mechanism of Injury

Severe, direct, or indirect impacts that dislocate the tibial head and femoral condyles or push the patella laterally can cause a subluxation. A patellar subluxation involving a congenital joint dysplasia can occur merely from a minor impact or accidental movement.

Types

A subluxation of the femorotibial joint can occur in the sagittal plane (anterior or posterior subluxation in 75 percent of cases), in the frontal plane (medial or lateral subluxation in 20 percent of cases), or in a rotary position. Usually, the ligaments are ruptured or are torn from their bony insertions.

Usually, the patella subluxes laterally. A recurrent subluxation of the patella can occur because of dysplasia or post-traumatically after the knee joint trauma.

Clinical Considerations and Diagnosis

The subluxation is characterized by a malalignment or deformation of the joint and loss of normal joint movement. The area around the joint is painfully swollen. The symptoms are less severe in a partial subluxation.

Medical Management

First aid for an acute subluxation of the knee joint consists of transportation with temporary immobilization. The subsequent treatment is rapid reposition under general anesthesia. Afterward, treatment must protect the healing of the capsule and ligaments by a long period of immobilization. Surgery is indicated primarily to ensure sufficient repair of the ruptured ligaments and early physical therapy (see earlier discussion of rotation injuries with ligament injuries). Surgery is also indicated if the reduction is not complete, as in subluxation fractures, which require realignment and fixation of the fragments. Accompanying injuries of subluxations (blood vessel and nerve lesions) must be treated first. Arteriography might be indicated to evaluate the extent of the vascular damage.

Late Complication

Recurrent subluxation of the patella will most likely cause early arthritic changes to the femoropatellar joint. The gliding path of the patella can be surgically corrected by reinserting the tibial patellar ligament more medially, by tightening the medial retinaculum, and by incision of the lateral retinaculum (in children, with an open epiphyseal growth plate, following procedures established by Goldthwait, in adults following procedures established by Klapp-Denks or Krogius-Roux).

MENISCUS INJURIES

Mechanism of Injury

A ruptured meniscus is the most frequent internal lesion of the knee joint. It is caused mainly by an indirect twisting force. The force turns the body on an immobile foot, or, vice versa, turns the foot and lower limb against the body (as in skiing). These forces can cause a laceration or complete separation of an intact medial or lateral meniscus.

Types

Lesions of the meniscus are classified as longitudinal ruptures (dislocation of the medial fragment into the joint space), "bucket-handle" tears, transverse ruptures, combinations of transverse and longitudinal ruptures, and partial and complete separations of the meniscus from the capsule (see Figure 8–3). An injury

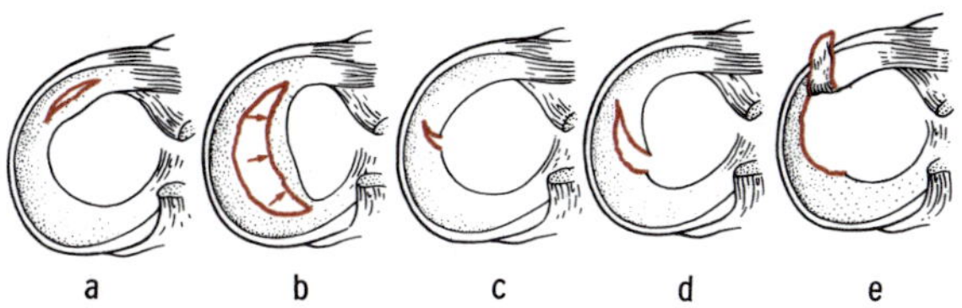

Figure 8–3 Meniscus lesions: a. longitudinal rupture; b. "bucket-handle" rupture; c and d. oblique ruptures; e. rupture with a flap

Source: Schlosser and Kuner, 1980

of the medial collateral lateral ligament can involve the meniscus because of the relatively close connection of the medial collateral ligament and the joint capsule adjacent to the medial meniscus. Such an injury might require arthroscopy to avoid diagnostic errors.

The meniscus can be completely torn even during normal activities if there are preexisting lesions, particularly a degenerative change of the meniscus itself. This is especially true of the lateral meniscus, which is torn much more frequently because of degenerative changes than a traumatic injury. A chronic knee injury can also cause meniscus degeneration and a rupture. This is a common injury among mineworkers.

Clinical Considerations and Diagnosis

The major symptoms of a meniscus lesion are pain with pressure to the joint space, inhibited extension, hyperextension, and rotation of the tibia with the knee flexed (Steinmann sign). In addition, capsular swelling and increased fluid (effusion in an acute meniscus rupture and hematoma with a capsular lesion) are observed, depending on the extent and duration of the lesion and the extent of previous degeneration. The acute injury is accompanied by severe pain. A patient suffering from a chronic injury often complains about a lack of stability. In "bucket-handle" ruptures of the meniscus, the soft tissue can be pinched. To exclude, in diagnosis, the possibility of loose joint bodies, radiograms are necessary.

Medical Management

Whether completely or only partially torn, the ruptured meniscus must be removed surgically. In surgery, the knee joint is examined, and possible injuries of the capsular ligaments or cartilage are evaluated and treated. Especially in chronic intra-articular lesions, there may be a proliferation of fatty tissue in the joint, which can cause persistent problems and require a partial resection.

In small lacerations of the meniscus and in bucket-handle ruptures, only the loose part is removed (partial meniscectomy). In young patients, if the meniscus protrudes from the capsule, it can be reinserted.

Late Complications

If for some reason a meniscectomy is not done, recurring damage of the meniscus and progressive joint disease will ensue. Even after a meniscectomy, degeneration of the affected joint compartment can develop, though considerably later.

Physical Therapy for Meniscectomy

Symptoms

The symptoms for a surgically treated meniscus rupture are the same as for a surgically treated rupture of the medial collateral ligaments, except that in the meniscus rupture the injured leg is usually wrapped in an elastic compression bandage.

Therapeutic Aspects

The therapeutic aspects are the same as in a suture repair of a lateral collateral ligament, except that in the case of a delayed healing modalities are added to procedures to improve circulation and the use of electrotherapy is added to procedures for strengthening.

Course of Treatment

Beginning on the second day, the patient exercises following the previously described guidelines for the prophylactic treatment of bedridden patients. In addition, the toe and foot muscles are contracted dynamically and the quadriceps muscle is contracted isometrically.

On the eighth day, the therapist adjusts the extension of the splint to approximately 15 degrees. The compression bandage is taken off, permitting dynamic knee exercises to begin. The most important exercises are those for full knee extension. At first, the flexion exercises are done very carefully to avoid an increased risk of effusion, which is common after this type of surgery. In some cases, edema is treated prophylactically with ice applications.

Beginning on the fifteenth day, the patient exercises in the therapeutic pool. The patient is allowed to ambulate with a knee splint and partial weightbearing with crutches using a three-point gait pattern. The amount of the weightbearing is adjusted to the limits of acceptable pain.

The rate of gradual withdrawal of supporting devices and aids depends on the functional and muscular improvement. Generally, if there is no effusion, the patient is able to walk within a week with a cane and without a splint. In the case of an effusion during any phase of the treatment, the procedures should be suspended and replaced by elevation, splint, isometric muscle contractions, ice, compression bandage, or possibly other modalities. Following the suspension, the exercises may be resumed, but with even greater care as to the rate of intensification.

After the eighteenth day, the range of motion is expanded and the strengthening and gait training are intensified. The patient also participates in swimming, occupational therapy, and recreational therapy.

Sample Exercises

The following exercises are used in the middle and late phases of strengthening:

1. with the splint on

    - actively moving the toes and ankle joint
    - isometric exercise of the quadriceps muscle including the foot, with (1) dorsiflexion and slight contraction of the middle head of the quadriceps, (2) dorsiflexion-eversion and slight contraction of the medial head of the quadriceps, and (3) dorsiflexion-inversion and slight contraction of the lateral head of the quadriceps
    - with approximately 30 degree knee flexion (by adjusting the splint, hand support, or foam pillow underneath the knee), dynamic contractions of the quadriceps muscle, including the foot as in the exercise above, with the lower leg partly supported initially by the therapist
    - elevating the leg in extension, including the foot as in the above exercise, with partial assistance initially

2. supine position

    - the same four exercises cited above
    - PNF: in a lateral meniscectomy, emphasizing the diagonal from adduction to abduction; in a medial meniscectomy, emphasizing the diagonal from abduction to adduction

3. sitting: PNF with knee extension and flexion
4. balance board

    - maintain balance
    - moving in different directions, holding, moving again

CARTILAGE AND BONE INJURIES

Mechanism of Injury

In a knee trauma, in addition to fractures of the joint-forming bones, lesions of the cartilage and bone may occur. Bone and cartilage fragments can be separated by direct or shearing forces.

Types

The injury might cause the formation of bony incongruities, and bone cartilage fragments can cause defects of the articular surface. The loose body fragments damage tissues in the same way as in a meniscus lesion.

Diagnosis

The diagnosis is by radiographic examination.

Medical Management

The small loose bodies are removed and the chronic condition eliminated. In an acute trauma, if the size of the fragment is large enough, reposition and refixation of the fragment are recommended. If the articular congruency cannot be restored in that way, the defect must be filled with autologous (from the posterior condyle) or homologous bone transplants. Smaller defects are drilled open, then filled gradually by medullary bone tissue.

Late Complications

Degenerative arthritis may occur as a result of incongruency.

Chapter 9

Injuries of the Tibia

TIBIAL FRACTURES

Mechanism of Injury

Fractures of the proximal and distal tibia are frequently caused by direct impact, especially in a longitudinal direction. Typically, fractures of the tibial shaft are caused by shearing, bending, compressing, or twisting forces (as in skiing accidents).

Types

Tibial fractures are classified as fractures of the proximal tibia, tibial shaft, and distal tibia (see Figure 9–1). Because of their special importance, malleolar fractures are discussed in the following chapter.

Fractures of the shaft usually involve both tibia and fibula; sometimes they involve the tibia alone. Occasionally, a direct impact causes an isolated fracture of the fibula, which is not clinically significant.

Clinical Considerations and Diagnosis

Typical symptoms are crepitation, swelling, and pain; frequently there are also skin contusions. Open (compound) fractures are common from traffic accidents. Simultaneous injuries of nerves (peroneal nerve) and blood vessels are possible; these require careful examination and observation and compressive bandages must be avoided. Radiographic pictures will reveal the type, position, and severity of the fracture.

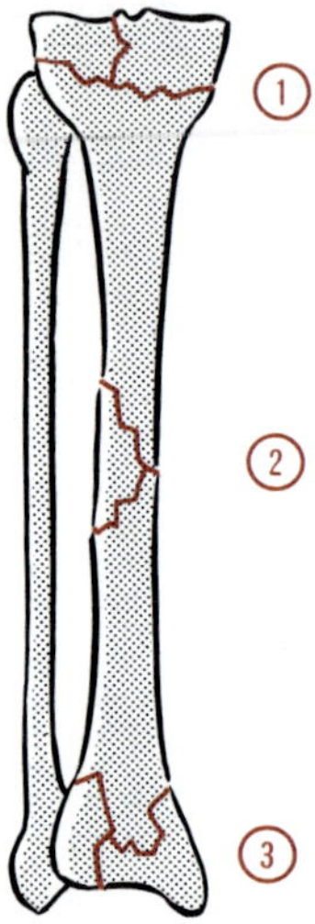

Figure 9–1 Tibial fractures: a. tibial head, b. tibial shaft, c. distal end

Complications

Hemorrhages and edema around the tibialis anterior, extensor hallucis longus, and extensor digitorum communis muscles result in ischemia, muscle necroses, and, subsequently, in contractures. This is called an anterior compartment syndrome. The treatment of choice is a surgical decompression: splitting the fascia, evacuating the hematoma, draining fluid, and so on.

PROXIMAL TIBIAL FRACTURES

Mechanism of Injury

A direct impact on the tibial head can produce longitudinal compression of the leg and a pulling on the ligaments and tendons, resulting in fractures of various types.

Types

Fractures of the spongy bone of the tibia occur in different shapes: condylar fractures, T-fractures, compression and comminuted fractures, and avulsion fractures involving the insertion of the ligaments.

Clinical Considerations

Fractures of the tibial head frequently affect the tibial joint surface and the ligamentous apparatus of the knee joint; they are usually combined with a hematoma in the knee joint. Occasionally, they are combined with fractures of the femoral condyles.

Medical Management

In undisplaced fractures, particularly in older patients, nonsurgical treatment is applied and splints can be used. The therapeutic goal in dislocated fractures is the restoration of an even joint surface, which requires surgery. Avulsed fragments with ligamentous insertions are screwed back in place, and compression fractures and fractures that interfere with the continuity of the joint surface are elevated and supported with spongy bone and then stabilized with screws or buttress plates. Exercises should start before the bony healing is completed, which takes about 8 to 12 weeks.

Early and Late Complications

In simultaneous fractures of the fibular head or lesion of the tibiofibular syndesmosis, the peroneal nerve can be affected. Uneven articular surfaces can result in arthroses.

Physical Therapy for Surgically Treated Fracture of the Tibial Head

Symptoms

The symptoms in a surgically treated fracture of the tibial head are the same as in an operatively treated lesion of the medial collateral ligament of the knee joint.

Therapeutic Aspects

The therapeutic aspects are similar to those listed for a suture repair of the medial collateral ligaments of the knee. An assistive dynamic splint (Frankfurt design) is used to complement mobilization.

Course of Treatment

In the first ten postoperative days, the patient exercises following the previously described guidelines for prophylactic therapy for bedridden patients. After about the eleventh day, the night splint need be worn only during the night. During the day, the patient works independently in a dynamic splint, which is temporarily removed during physical therapy. The entire leg is mobilized and strengthened, at first without weightbearing. Special emphasis is placed on exercises for the quadriceps muscle. If there is a tendency to have effusion, which is frequently the case in these fractures, ice therapy should be applied.

After about the fifth week, exercises in the therapeutic pool are added to the treatment. One or two weeks later, the patient can wear a long elastic stocking and is allowed to walk with crutches with no or minimal weightbearing (touch-toe contact).

Beginning in the thirteenth week, if the fracture is sufficiently consolidated, the patient can increase weightbearing gradually; the rate of increase depends on the individual circumstances. A correct gait must be practiced consistently. Recreation therapy and occupational therapy are additional therapeutic possibilities.

Sample Exercises

The following exercises are used in the early, middle, and late phases of mobilization and strengthening. The first set is done with the patient in a dynamic splint (Frankfurt design):

- active toe and foot exercises
- isometric contractions of the quadriceps muscle (including the ankle)
- combined flexion/extension of knee and hip: holding at the end of the flexion or extension, relaxing completely, then flexing/extending a little more, making sure the normal leg remains on the bed
- with the therapist holding the splinted knee in extension, active knee flexion as far as possible, then elevating in extension as far as possible (both exercises should include ankle exercises)
- with knee splint and leg in traction, hip abduction with the best possible hip extension and adduction back to the midline

Next, with the patient sitting on a stool, the injured leg is extended beside the normal leg (flexed 90 degrees). Then the heel is pulled gently along the floor under the stool for knee flexion with ankle dorsiflexion, then brought gently forward again into extension as far as possible.

Finally, the patient sits on a gymnastic ball and rolls backward and forward as far as possible, with the injured leg assisted by the normal leg in bilateral symmetrical exercises.

Paralysis of the Peroneal Nerve

Therapeutic Aspects

For the treatment of a paralyzed peroneal nerve, the following aspects are important:

- prevention of pressure sores
- prevention of contractures with an overnight splint (Figure 9–2)
- prevention of equinus by passive exercises for dorsiflexion (teach to do)
- circulatory support by hot and cold baths for the foot
- electrotherapy, as long as there is no metal implanted material in the electrical field
- muscle reeducation and strengthening exercises, if possible with PNF
- gait training

For details of the treatment for a paralyzed peroneal nerve, a neurologic textbook should be consulted.

Sample Exercises

The following exercises are used in the early, middle, and late phases of contracture prevention and as strengthening exercises. First, while in a supine position or sitting with knees extended, the patient pulls the ankle, assisted by tubing, into dorsiflexion, dorsiflexion/inversion, and dorsiflexion/eversion.

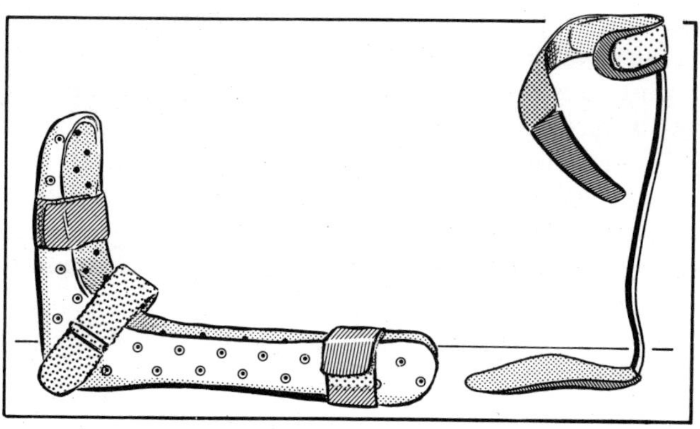

a b

The following set of exercises is done with the patient in a supine position:

- using PNF, adduction diagonal, exercising the tibialis anterior by a quick stretch and maximum resistance to the proximal components
- using PNF, abduction diagonal, exercising the peroneal muscles by a quick stretch and maximum resistance to the proximal components
- bilateral symmetrical exercises of all ankle and toe movements
- extension of the big toe, without the other toes
- extension of the small toes, without the big toe

Finally, while sitting on a stool with knee flexed at a right angle and feet on the floor, the patient symmetrically brings both ankles into dorsiflexion, relaxes, then into inversion and eversion.

Before each of the preceding exercises stimulation by quick icing may be beneficial in activating a muscle contraction.

TIBIAL SHAFT FRACTURES

Mechanism of Injury

Tibial shaft fractures are caused by a direct impact or, indirectly, by twisting and bending forces.

Types

Fractures of the tibial shaft usually occur in combination with fibular fractures. With fractures of the tibia alone, the intact fibula interferes with the realignment of the tibial fragments, which results in delayed healing and an increased risk of pseudoarthrosis. Compound fractures of the lower tibia are frequently complicated by soft-tissue injuries and considerable malalignment.

The following types of fractures can occur: transverse and oblique fractures, bending fractures with dislocated wedges, torsion fractures, and comminuted fractures with several or multiple fragments. In torsion fractures caused by indirect force (rotation of the body with the foot fixed), the fractures are usually in the distal third of the tibia and the proximal third of the fibula.

Clinical Considerations

Fractures of the tibial shaft produce the typical symptoms of a fracture: deformity, dislocation, and crepitation.

Medical Management

The type of treatment depends on the location of the fracture, the situation of the surrounding soft tissue, and the age and general condition of the patient. If nonsurgical and surgical treatments appear to be equally likely to result in restoration, the nonsurgical treatment is preferred, and is in fact frequently selected for fractures in adolescents and/or for undisplaced fractures. Immobilization in a cast is required for 6 to 12 weeks. In a dislocated fracture, open reduction and internal fixation of the fracture and subsequent immobilization in a cast are recommended. However, if the fragments cannot be reduced and retained properly, if compressed tissue between the fragments is suspected and/or severe skin lesions are present, an osteosynthesis is preferred to avoid complications and nonunion of the fracture.

The medullary nail (rod) has proved to be successful in osteosyntheses of transverse or short oblique fractures. In other types, plates and screws are used. Compound and comminuted fractures can also be stabilized by external fixation devices.

Usually, excellent adaptation of the fragments and a good bony stability for exercise can be achieved by an osteosynthesis, thereby avoiding the problems of a long immobilization period.

Early and Late Complications

After nonsurgical treatment, malalignment and pseudarthroses can occur. Also, an insufficient and unstable osteosynthesis can result in a pseudarthroses and infection.

Physical Therapy for Surgically Treated Fracture of the Tibial Shaft (Plate-Screw Osteosynthesis)

Symptoms

With the injured leg bandaged and placed in a splint, the following symptoms may be observed:

- postoperative edema
- generally decreased activity due to pain and position, also decreased respiratory and circulatory functions
- pain around the fracture site

Therapeutic Aspects

- pressure area prophylaxis: (1) isometric and dynamic exercises, (2) changing positions
- pneumonia and thrombosis prophylaxis: (1) respiratory therapy, (2) isometric and dynamic exercises
- circulatory stimulation: (1) active movement, (2) changing positions, (3) therapeutic pool, (4) cyrotherapy
- edema prophylaxis: (1) elevation, (2) compression, (3) isometric and dynamic muscle contractions
- mobilization: (1) isometric and dynamic exercises, (2) relaxation techniques with ice, (3) active movement with equipment, (4) therapeutic pool, whirlpool
- strengthening: (1) dynamic exercises, (2) exercises against resistance, with or without equipment, (3) therapeutic pool, (4) recreation therapy
- gait training

Course of Treatment

Beginning on the first postoperative day, the patient exercises following the previously described guidelines for prophylactic treatment of bedridden patients. The patient is allowed to do isometric and active exercises with the injured leg. During the first few days, it might be necessary to support the weight of the leg.

After the eleventh day, if the healing progresses without complications, the treatment is supplemented by hydrotherapy. The patient is allowed to walk with crutches, either nonweightbearing or with toe touch only. During upright activities, the patient should wear an elastic stocking to avoid edema; if it does occur, edema treatment must be started (elevation, and so on).

In the tenth week, normal weightbearing is begun, with the weight increased as indicated by radiographic examination. Now, the main components of the therapeutic program are exercises against resistance and from partial weightbearing positions, exercises while standing upright or on a balance board, and gait training.

Frequently, the patient has difficulty putting weight on the injured leg either because of anxiety or because of pain. Thus, to reassure the patient and facilitate weightbearing, the therapist should regularly examine with the patient appropriate weightbearing on the scales. The therapist should also train the patient to shift weight during the appropriate phases of gait (perhaps using a mirror).

In addition to the regular physical therapy program, the patient engages in recreational therapy.

Sample Exercises

The following exercises are used in the middle phase of edema prophylaxis to encourage circulation. The first set of exercises involves changing positions on an elevated table (see Figure 9–3). The patient lies with elevated legs and exercises ankle and toe joints, with alternating movements (approximately ten times each). Immediately afterward, in the supine position or sitting on the edge of the bench, the patient alternates intensive movement of toes and ankle with isometric contractions of all leg muscles (again ten times each). The two exercises should be repeated three times, finishing in the supine/sitting position. The application of ice packs can also prevent edema.

The second set of exercises employs a foam block or ball. The patient lies in a supine or semi-sitting position. The patient holds the ball between the feet and against the therapist's attempts to take it away. The resistance should be minimal and isometric contractions should be the reaction; excessive shearing forces on the fracture should be avoided. Then, in the same position, the patient engages in the widest possible range of active knee and hip joint movements.

Physical Therapy for Surgically Treated Fracture of the Tibial Shaft (Intramedullary Nailing)

Symptoms and Therapeutic Aspects

The symptoms and therapeutic aspects are the same as those for surgically treated fractures of the tibial shaft (plate-screws osteosynthesis).

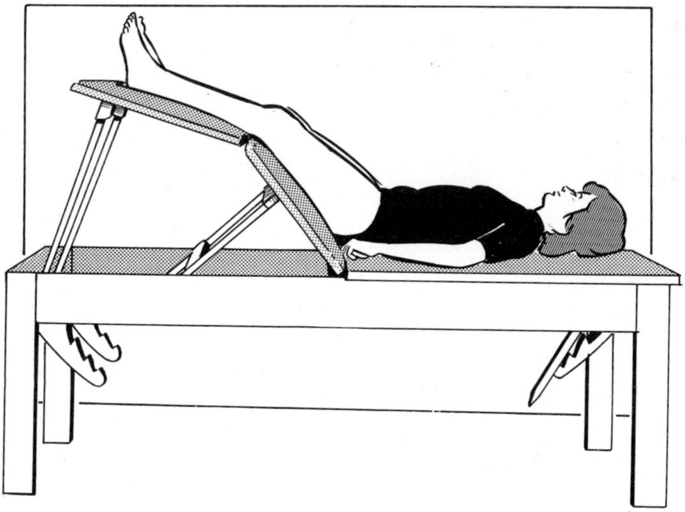

Figure 9–3 Elevated table support

Course of Treatment

On the first postoperative day, the patient begins exercises following previously described guidelines for the prophylactic treatment of bedridden patients, supplemented by dynamic exercises of the entire leg. Initially, the leg must be supported during hip and knee exercises.

After the third day, hip and knee exercises against resistance begin. However, the patient must avoid any rotary or shearing stress on the lower leg. After ten days, if the wound has healed completely, the patient is allowed to enter the therapeutic pool. Contrast foot baths or whirlpool may be beneficial.

Walking with crutches is permitted. Usually, this starts with partial weightbearing of 20 kg and is then gradually increased in the course of treatment, depending on individual ability. Only after the fracture has been completely consolidated may exercises with rotary and shearing stress be started.

Sample Exercises

The following exercises may be used in the early, middle, and late phases of strengthening and gait training, using air pillows (see Figure 9–4):

- while lying in a supine position in bed, alternate stepping motions against the air pillows by pushing the feet fully against the pillows

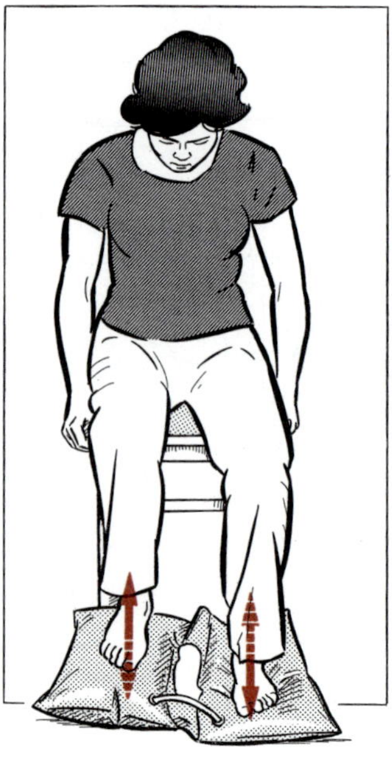

Figure 9–4 Air pillow

- doing the same exercise while sitting on a stool
- while standing between parallel bars, first doing the above exercise, then balancing while standing at the middle of the two pillows

DISTAL TIBIAL FRACTURES

Fractures of the distal tibia (excluding malleolar fractures) are similar to proximal fractures with respect to their origin, consequences, and therapeutic aspects. A tibial fracture that involves the articular surface of the talus is of particular importance (see Figure 9–5).

Types

Weber classifies these fractures into three types:

Type A: separation of anterior and posterior fragments and severe compression of the spongy bone
Type B: shearing of a large anterior fragment
Type C: shearing of a large posterior fragment

Not infrequently in infants and adolescents, an injury of the distal epiphyseal plate occurs instead of a distal fracture. In combination with these traumatic epiphysiolyses, meta- and/or epiphyseal tibial wedges can be split off (see Part I).

Medical Management

Surgical treatment must be employed. Since the uneven surfaces can cause joint degeneration, a careful restitution of the anatomical structures is required. Defects

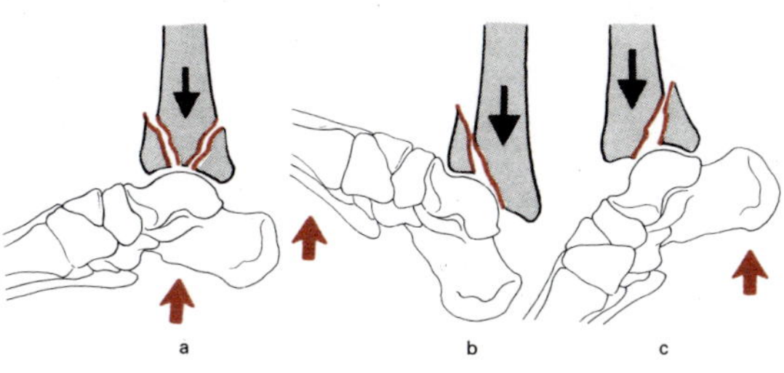

Figure 9–5 Types of fractures of the distal tibia: a. separation of anterior and posterior margins; b. shearing of the anterior margin; c. shearing of the posterior margin

Source: Schlosser and Kuner, 1980

are filled with autologous spongy bone, and the fixation is achieved by buttress plates and screws. Afterward, a cast is applied, permitting dorsiflexion exercises.

Early and Late Complications

The uneven articular surfaces can result in joint degeneration. The resulting increasingly painful range of motion restrictions require orthopedic shoes and, possibly, the immobilization of the joint by an arthrodesis.

Chapter 10

Injuries of the Tarsus, Metatarsus, and Phalanges

TARSAL JOINT INJURIES

The functional unit of the talocrual, proximal, and distal tarsal joints have major clinical significance. These joints are in a very peripheral position with respect to the loading axis of the body. With the foot fixed on the ground, the resulting forces of the long lever arms subject these joints to numerous injuries. The most frequent injuries are ligament injuries and bone fractures in the area of the tarsal joints. Their severity is not always recognized, however, and the diagnosis and treatment of the ruptured ligaments are often inappropriate. The ligaments also play an important role in injuries and management of malleolar fractures.

Because of the physiological position of the hind foot valgus, the lateral part of the joint is subjected to more weightbearing than the medial part. As a counterpart to the talus, the fibula tolerates loads as heavy as 50 percent of the body weight. The importance of an intact syndesmosis between the tibia and the fibula as a basis for restoration after an injury cannot be emphasized too strongly.

Tarsal injuries are characterized by pain with palpation or weightbearing and considerable edema due to hemorrhage and inflammation. This must be taken into account, especially when applying supporting and immobilizing bandages to prevent decreased blood flow by tight elastic bandages.

LIGAMENT INJURIES OF THE PROXIMAL TARSAL JOINT

Mechanism of Injury

Contusions, twisting, and overstretching as a result of excessive ankle movements occur especially during sports, but they can also occur in daily activities. The foot is in inversion and internal rotation, or less commonly in eversion and external rotation, and a sudden and forceful stress causes the joint to go beyond its normal physiological range of motion, thus resulting in ligamentous sprain.

Types

As with the knee joint, torsion, overstretching, and rupture must be distinguished. The tibia and fibula are connected by a syndesmosis and ligaments. During extreme dorsiflexion of the ankle, the talus is pushed between the malleoli, which may result in overstretching the ligaments and lead to disruption of the syndesmosis. Excessive plantarflexion can result in a passive overstretching of the anterior joint capsule and, in case of repeated stress, in arthrotic spurs from the anterior edge of the tibia. By exceeding the range of motion for eversion or inversion, overstretching and rupture of the respective medial or lateral ligament can result. Ligament tears and ruptures occur mainly in the lateral ligaments of the ankle joint.

Clinical Considerations and Diagnosis

The symptoms are pain on palpation, edema, hematoma, pain when stressing the joint in inversion or eversion, and increased opening of the joint space. If the anterior structures are affected, a lateral radiograph will show an anterior subluxation of the talus against the tibia. Disruptions of the syndesmosis are radiographically characterized by a widening of the space between the medial and lateral malleolus and the syndesmotic gap. They frequently occur with small split bone fragments.

Medical Management

The type of treatment depends on the extent of the injury. Nonsurgical treatment is indicated when the capsule or ligaments are not ruptured; for example, an overstretching injury may be treated by the application of a cast to the lower leg for 4 to 6 weeks. Surgery is indicated for ruptures of the capsule and ligaments, with the application of a cast for 4 to 6 weeks. In subluxations of the proximal tarsal joint, the tibia and fibula are usually involved. The talus can sublux into an anterior, posterior, or lateral direction. A subluxation is always combined with a partial or complete rupture of ligaments, which requires a surgical suture repair.

Early and Late Complications

An unstable ligament with recurrent injury and degeneration must be treated with surgical repair of the ligament. Care must be taken to avoid ligament insufficiency and necrosis of the talus.

MALLEOLAR FRACTURES

Mechanism of Injury

In malleolar fractures, the fracture fragments are usually broken off and the ligamentous insertions are torn out with an adjacent bone chip. The traumatic mechanism is usually similar to that of ligament ruptures.

Types

The typical malleolar fracture is frequently an eversion fracture caused by forceful abduction of the proximal talar joint and, less often, an inversion fracture caused by forceful adduction of the talus. The inversion trauma is associated with adduction and inversion of the foot; the eversion trauma is associated with an eversion and external rotation. In an inversion injury, the medial malleolus fractures medially, and the lateral ligaments or the tip of the lateral malleolus may be avulsed. In an eversion injury, the ligamentous rupture or avulsion of the malleolus occurs medially, and the fibula fractures laterally.

The three-type classification by Weber is shown in Figure 10–1. In this classification schema, the type of malleolar fracture is determined chiefly by the ligaments of the tarsal joint. Locale of the fibular fracture is also an important factor. The three types are:

Type A: The fracture is below the syndesmosis, which is intact.
Type B: The fracture is at the level of the syndesmosis, which is partially ruptured.
Type C: The fracture is above the syndesmosis, and there is a lateral subluxation.

The posterior edge of the tibia can also be split off.

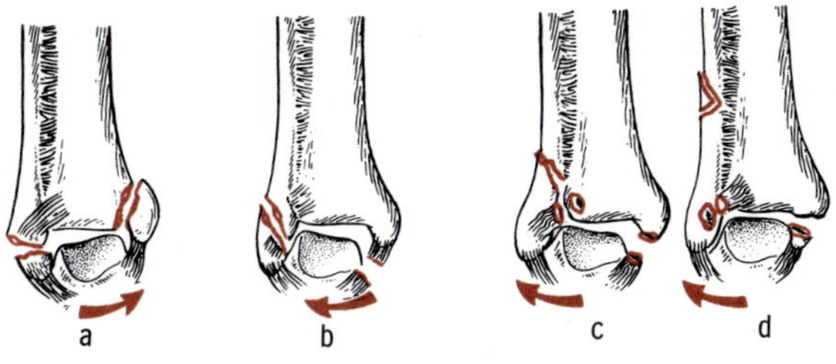

Figure 10–1 Injuries of the talocrual joint: a. inversion injuries; b. eversion injuries (Weber Type A); c. injury above the syndesmosis (Weber Type C); d. Weber Type C with higher fibula fracture

Source: Danis, 1947; Willenegger and Weber, 1963

Symptoms, Clinical Considerations, and Diagnosis

The symptoms of a malleolar fracture are edema, external deformation, palpable bony deformation, and loss of physiological range of motion. The diagnosis is verified by radiographic examination. A considerable amount of malalignment and edema may interfere with the blood flow and peripheral nerves.

Medical Management

The anatomical structures must be restored without the slightest irregularity and the ligaments, including the syndesmosis, must be repaired. Type A fractures are treated nonsurgically in a cast for about six weeks. Type B and C fractures have to be treated surgically. The shape and length of the skeletal components are restored exactly as before and stabilized by screws or plates. The ligaments are sutured. The syndesmosis is stabilized temporarily by a screw, which has to be removed before partial weightbearing is started. With multiple lesions, immediate elimination of the malalignment and reduction of a subluxation fracture is a priority. If the osteosynthesis does not produce sufficient stability for exercise, a brace worn for several weeks is indicated.

Complication

A possible complication is degeneration due to uneven articular surfaces.

Physical Therapy For Suture Repair of the Lateral Collateral Ligament of the Proximal Tarsal Joint

Symptoms

With the injured leg supported by a bivalved cast, the following symptoms are observable:

- postoperative edema, possibly also hematoma
- possibly reduced general activity, resulting in decreased circulation
- pain around the surgical incision

Therapeutic Aspects

- edema prophylaxis: (1) elevation, (2) compression, (3) isometric and dynamic exercises

- circulatory stimulation: (1) active movement, (2) changing positions, (3) cryotherapy, (4) hydrotherapy
- mobilization: (1) isometric and dynamic exercises, (2) relaxation techniques with ice, (3) active exercises with exercise equipment, (4) hydrotherapy
- strengthening: (1) dynamic exercises, (2) resistive exercises with or without equipment, (3) recreation therapy
- training in gait activities

Course of Treatment

During the first three postoperative days, the bivalved cast remains closed. The therapist works with the patient following the previously described guidelines for edema prophylaxis. On the fourth postoperative day, the therapist opens the cast and instructs the patient to dorsiflex the ankle. After the exercises, the cast is secured by an elastic bandage. The patient is allowed to exercise carefully doing dorsiflexion and plantarflexion, and, if possible, inversion and eversion. On the tenth day, if the wound is healing well, a walking cast is applied and it is then worn for about four weeks.

Following postoperative immobilization of six weeks, the walking cast is removed and the main therapeutic goals are the mobilization of the joints and a strengthening of the muscles, without restrictions and against controlled resistance. The patient also receives hydrotherapy.

Although full weightbearing is permitted, this option is usually precluded by continued swelling, joint range of motion limitations, and pain. Intensive walking exercises are started, adjusted to individual needs and directed at achieving a physiological gait without assistive devices. The treatment after a surgical repair of ligaments using synthetic fibers is similar to that applied after the suture repair of a ligament of the tarsal joint.

Sample Exercises

The following exercises are used in the middle to late phases of mobilization, strengthening, and gait training. In the first set of exercises, the patient sits on a stool and uses both feet and a plastic bag:

- folding and unfolding the bag
- rolling the bag together and flattening it back onto the floor
- turning the bag inside out
- crumpling the bag

The following set of exercises are done standing:

- stepping forward with one leg and alternately shifting the weight forward and backward
- with the lower extremities in abduction, shifting the weight alternately side to side

SURGICALLY TREATED FRACTURES OF THE PROXIMAL TARSAL JOINT

Symptoms, Therapeutic Aspects, and Course of Treatment

The symptoms, therapeutic aspects, and course of treatment up to the third postoperative day are to a great extent the same as for suture repair of the lateral collateral ligament of the proximal tarsal joint. On the fourth postoperative day, the cast is opened, and dorsiflexion of the ankle can be tried carefully. On the eleventh day, if the primary wound is healing, a walking cast is applied. It is worn for 4 to 5 weeks, depending on the results of the radiographic examination.

In the seventh week, the cast is removed. Radiographic examination will help to decide whether another cast has to be applied until the twelfth week or whether an active exercise program can be started. If exercises are possible, edema control is indicated in most cases. Exercises include those to increase the range of motion of the proximal and distal tarsal joints and to strengthen all the leg muscles. Hydrotherapy is then added to the program. Walking without weightbearing or with toe-touch weightbearing on crutches is allowed, with the patient wearing a compression stocking.

Beginning in the thirteenth week, the patient is allowed to bear 10 to 20 kg of weight. The weightbearing is increased gradually, depending on the individual situation. Training in posture and gait can now be added to the program.

Sample Exercises

The following exercises are used in the middle and late phases of mobilization, strengthening, and gait exercises:

1. sitting on a stool

    - The knees are extended with feet resting on the heels while performing all toe and ankle movements, bilaterally and symmetrically.

- The feet are placed on the floor with the knees at 90 degrees of flexion; dorsiflexion and plantarflexion is done alternately, bilaterally, and symmetrically.

2. While sitting on a gymnastic ball with the feet on the floor, patient rolls the ball forward, backward, to both sides, and in a circle (both directions), keeping the feet firmly on the floor.
3. balance board

- shifting weight forward and backward
- shifting weight to the sides

ACHILLES TENDON RUPTURE

Mechanism of Injury

A rupture of the Achilles tendon is only rarely the result of an external injury (blow, cut, stabbing). Usually, it is an endogenous injury caused by an indirect force (skiing or other athletic activity) or it is caused spontaneously as a result of preexisting degenerative lesions or cortisone injections. In certain circumstances, the rupture may be triggered by an abrupt change of muscle tension.

Types

Ruptures of the tendinous portion close to the calcaneus are usually the result of degenerative changes of the tendon and the tendon sheath. Ruptures at the transition of muscle and tendon with an adjacent bone chip are caused by a trauma. These can occur during extreme forceful overstretching, for example, from an impact against the dorsiflexed flexed foot while skiing.

Degenerative changes of the tendon and tendon sheath are the result of age-related processes and/or local malnourishment combined with excessive use. A pending rupture can be indicated by achillodynia, which is a painful irritation of the tendon.

Clinical Considerations and Diagnosis

Normally, a partially ruptured Achilles tendon does not cause severe pain. In a fresh injury, a transverse depression is palpable; the depression is filled by a hematoma and edema and then disappears. Plantar flexion of the foot is absent; standing on tiptoes with the injured leg is not impossible, with a complete rupture. The intact tendon of the plantaris muscle maintains some function, but is not strong enough to perform standing on the toes. The reflex of the Achilles tendon is absent.

Medical Management

The indicated treatment of a ruptured Achilles tendon is surgical repair by single or continuous sutures. The plantaris tendon can also be subject to surgical repair. In more severe degenerative lesions and dislocations, a plastic repair must be considered. A bone fragment that is torn off at the tendonious insertion is fixed back into place by screws. After an immobilization period of three weeks, a walking cast is applied for 3 to 5 weeks. Usually, exercises can be started after 6 to 8 weeks. During the following months, an elevated heel is recommended to partially relieve the weightbearing stresses to the Achilles tendon.

Physical Therapy

The following therapeutic program is applicable to a suture of the Achilles tendon.

Symptoms

The injured leg is in a bivalved cast with the proximal tarsal joint in an equinus position to relax the tendon and suture repair. The following symptoms are observable:

- postoperative edema, clinically not obvious
- decreased general activity, resulting in decreased circulation
- pain around the injury

Therapeutic Aspects

- edema prophylaxis: (1) elevation and compression, (2) isometric and dynamic exercises
- circulatory stimulation: (1) active movement, (2) isometric and dynamic exercises, (3) hydrotherapy
- mobilization: (1) isometric and dynamic exercises, (2) possibly relaxation techniques with ice, (3) hydrotherapy
- strengthening: (1) dynamic exercises, (2) resistive exercises, with or without equipment
- gait training

Course of Treatment

During the first ten postoperative days, the therapist works with the patient following previously described guidelines for prophylactic therapy for edema. After the tenth day, a walking cast is applied.

In the seventh week, following removal of the cast, hydrotherapy mobilization, and strengthening of the ankle are begun. Exercises are allowed in all positions and in all directions. However, initially dorsiflexion should be done only actively, without resistance and weightbearing.

Walking at first requires crutches with only partial weightbearing, with the patient wearing compression stockings and with heels elevated 3 cm. The patient should try as soon as possible to walk with only one cane and then, finally, without any assistive devices. At the end of each month, the heel is lowered one centimeter, enabling the patient to regain the normal height of the heel after three months. At that point, the patient should be able to walk barefoot with full weightbearing.

Sample Exercises

The following exercises may be used in the middle phase of the mobilization and strengthening. First, with the patient in a supine position, all movements of the ankle, especially dorsiflexion, are done actively and without resistance.

Next, with the patient sitting on a stool, the feet on the floor and knee joints flexed in an obtuse angle, the heels are alternately elevated. The knee flexion is then gradually changed to a right angle. Using rings placed on the floor, and with the ankle dorsiflexed, the patient then tries to put the heel into each ring. Finally, a rod is placed on the floor behind the patient's heels, with knee joints flexed in an obtuse angle. The patient tries to put the dorsiflexed foot flat on the floor behind the rod and then again in front of the rod.

In a third set of exercises, the patient stands without weightbearing on the injured leg while holding on to a rail, then (1) rolls a ball under the foot, (2) places the leg with the ankle dorsiflexed on a box and then back down on the floor, and (3) tries to put the full sole of the foot flat onto the floor.

FOOT INJURIES

Subluxation and Sprain

The subluxation of the talus, which usually occurs in conjunction with lesions of the articulating components of the fibula and tibia, was discussed earlier in the section on malleolar fractures. A subluxation of the distal tarsal joint always includes a subluxation of the talonavicular joint and is accompanied by severe injuries of the ligaments, sometimes even of the bone.

Mechanism of Injury

This subluxation may be caused either by a direct impact, with fixed foot, or by an indirect force.

Types

The subluxation occurs primarily in a medial direction (50 percent of cases), but it can also occur in lateral, dorsal, and plantar directions.

Clinical Considerations and Diagnosis

The subluxation produces deformation and pain. The diagnosis is confirmed by radiograms.

Therapy

The subluxation must be reduced immediately because of the decreased blood flow. This is followed by immobilization in a cast for 4 to 6 weeks.

Frequently, surgery is indicated because of constricted circulation to the tissue. A reposition and fixation with a Kirschner wire can restore ligament stability and lay the basis for early exercises. Subluxations of the joints of the metatarsal and small tarsal joints can occur in conjunction with fractures of the base of the metatarsal bones. These are discussed in a later section.

Sprains of the toe joints are immobilized temporarily with adhesive bandages. These sprains do not interfere with the ability to stand and walk. Subluxations of the toe joints are reduced and treated with a spica bandage for a short period of time.

TALUS FRACTURE

Mechanism of Injury

Fractures of the talus are caused by a strong direct impact.

Types

Talus fractures are classified as central fractures of the body, neck, and head (peripheral fractures with fragments torn from the lateral edge) and fractures of the lateral and posterior processes of the talus (see Figure 10–2). These can be either

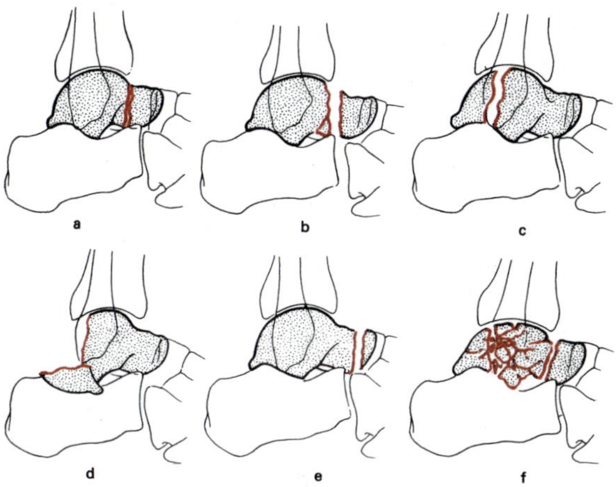

Figure 10–2 Types of talus fractures: a. fracture of the neck; b. displaced fracture of the neck; c. fracture of the body; d. subluxed fracture of the body; e. fracture of the head; f. comminuted fracture

Source: Schlosser and Kuner, 1980

contusion or comminuted fractures. The fragments can be dislocated, and bony ridges can interfere with the articular congruency of the distal talar joint. The small tarsal bones can be subluxated against the tarsus.

Clinical Considerations and Diagnosis

The symptoms are painfully limited ankle range of motion, deformation, and palpable ridges. Radiograms in two planes are used to verify the type of fracture.

Medical Management

For peripheral and central fractures, which can be precisely reduced, the treatment is nonsurgical. Active functional exercises are started after immobilization in a cast for four weeks. There is no weightbearing for 12 weeks. Severely dislocated fractures have to be reduced surgically and stabilized with screws. A completely even alignment must be achieved.

Fractures of the body and neck and subluxation fractures require a long period of nonweightbearing because of the frequently interrupted blood circulation; this may require approximately one year in a specialized walking orthosis.

Early and Late Complications

Fractures of the body and neck of the talus, especially subluxation fractures, can cause necrosis of the talus because of the disturbed blood supply. Remaining bony ridges can result in a post-traumatic degeneration of the talus and affect nerves and vascular tissue.

Physical Therapy for Nonsurgical Treatment of Talus Fractures

During the approximately four weeks of immobilization in a cast, the patient continues all activities of daily living not restricted by the casted ankle and foot. After the removal of the cast, the following physical therapy is scheduled:

- edema prophylaxis
- stimulation of circulation
- mobilization
- strengthening

Weightbearing is normally possible after 12 weeks. At that point walking exercises may also begin.

CALCANEUS FRACTURE

Mechanism of Injury

Fractures of the calcaneus usually occur in the area adjacent to the talus. They are caused primarily by compression, for example, in falling from a great height. Frequently, they occur bilaterally and result in a considerable deformation of the calcaneus, that is, a flattening of the crucial angle within the subtalar joint.

Types

In addition to comminuted compression fractures (see Figure 10–3), isolated oblique and transverse fractures can occur. Fractures with avulsion of the sustentaculum tali or the Achilles tendon occur less frequently (see Figure 10–4). According to Vidal, three categories of severity may be distinguished:

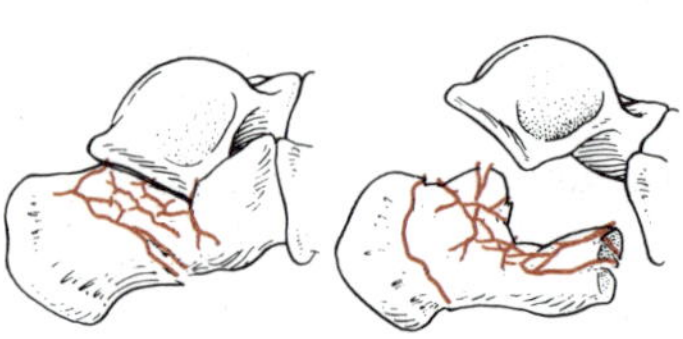

a b

Figure 10–3 Fractures of the calcaneus: a. compression fracture; b. comminuted fracture

Source: Schlosser and Kuner, 1980

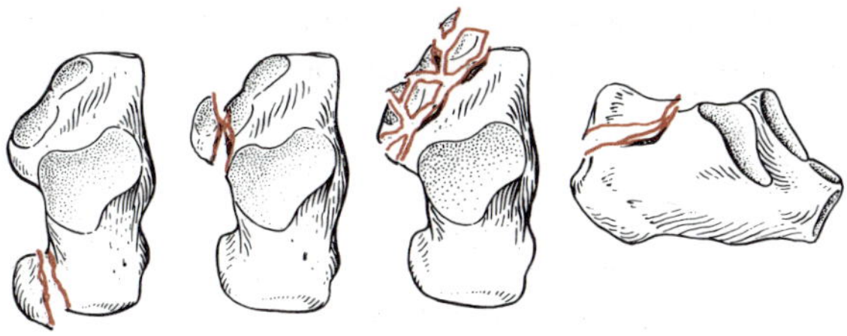

Figure 10–4 Fractures of the calcaneus: a. separation fracture of the tuberosity; b. sustentaculum tali separation; c. medial comminuted fracture without joint lesion; d. avulsion with Achilles tendon

Source: Schlosser and Kuner, 1980

---

1. an isolated fracture without affecting other joint structures (9 percent of cases)
2. a comminuted fracture involving other joint structures to some degree (27 percent of cases)
3. a comminuted fracture combined with severe lesions to other joint structures (64 percent of cases)

Clinical Considerations and Diagnosis

Fractures of the calcaneus produce deformation, swelling, and hematoma around the posterior ankle. Radiographic examination is used to confirm the diagnosis.

Differential Diagnosis

A contusion of the calcaneus causes similar symptoms but no bone lesion. Like the fracture, it can be caused by falling or jumping from a great height and requires nonweightbearing for several weeks.

Medical Management

Category 1 and 2 fractures are treated nonsurgically by immobilization in a cast with the heel suspended for six weeks. Physical therapy begins without weightbearing, then progresses gradually with increased weightbearing. Category 3 fractures are treated by cast immobilization with surgery indicated only in select cases. Open reduction and internal fixation is not normally required, and an early arthrodesis is considered only in special cases. The separated fragments are fixed with screws. Usually an arch support, or even orthopedic shoes, are necessary.

Early and Late Complications

The range of motion of the subtalar joint may remain limited. Occasionally, subtalar arthrodesis has to be done to correct complications.

Nonsurgically Treated Fractures of the Calcaneus

Symptoms

The injured ankle and foot is supported by a cast for 10 days. After the eleventh day, the foot is placed in a splint. The following symptoms are observable:

- edema around the heel and the tarsal joints, sometimes also the forefoot
- discoloration from a residual hematoma around the heel, sometimes also toward the middle foot, and possibly taut shiny skin
- limited range of motion in the proximal and especially the distal tarsal joint due to edema and pain, sometimes also due to the position of the fracture
- possibly general inactivity and reduced circulation
- pain around the fracture site

Therapeutic Aspects

- edema proxphylaxis: (1) elevation and compression, (2) isometric and dynamic exercises, (3) ice application
- circulatory stimulation: (1) isometric and dynamic exercises, (2) ice applications, (3) foot baths
- mobilization: (1) dynamic exercises, (2) foot baths
- strengthening: (1) dynamic exercises, (2) resistive exercises with equipment
- gait training

Course of Treatment

In the first ten days, the therapist has the patient perform isometric and dynamic exercises for all body muscles that are not restricted by the cast. On the eleventh day, after the removal of the cast, the lower leg is bandaged or a compression stocking is applied, as well as a splint. The leg must be elevated in the splint and ice packs applied as long as the edema and hematoma have not subsided.

After 2 to 3 weeks, hematoma and swelling have usually subsided. Exercises are begun with full range of motion of the proximal and distal tarsal joint, and all leg muscles are strengthened.

The patient is now allowed to walk with crutches, nonweightbearing. Along with the mobilizing and strengthening exercises, the patient receives hydrotherapy and starts swimming.

In about the thirteenth week, after the patient has been supplied with an arch support or with orthopedic shoes and depending on the results of radiographic examination, weightbearing begins. The progress and rate of increase in weightbearing will depend on the individual situation. Usually, it requires time and effort to progress from partial weightbearing to full weightbearing with a normal gait pattern.

Sample Exercises

The following exercises are used in the middle and late phases of mobilization, strengthening, and gait training:

1. while sitting down with extended knees
    - active and resistive ankle and toe movements
2. while sitting with feet on the floor
    - crawling forward with the toes
    - grasping pebbles, small foam pieces, and pencils with the toes, then moving them from one spot to another
    - grasping a towel, rolling and folding it, lifting it up
    - with a small ball under the foot, rolling it around
3. in a foot bath
    - active ankle and toe movements

FRACTURES OF THE SMALL TARSAL BONES

Mechanism of Injury

Fractures of the tarsal bones are caused mainly by a direct impact, for example, by compression of the foot when it is run over by a car.

Types

In most cases, several tarsal bones are injured. The fractures are classified as simple fractures or comminuted fractures with several or multiple fragments.

Clinical Considerations and Diagnosis

The symptoms are edema, hematoma, and foot deformities, accompanied by pain on compression and palpation.

Medical Management

Simple fractures are immobilized in a cast. More severe fractures with separation have to be reduced surgically and stabilized with Kirschner wires. The postoperative immobilization period is 12 weeks.

Nonsurgically Treated Fractures of the Small Tarsal Bones

Symptoms

With the injured leg supported by a cast, the following symptoms are observable:

- considerable edema around the foot
- edema limiting range of motion of all foot joints
- pain and sometimes sensitive, ''cold'' toes
- general inactivity and reduced circulation

Therapeutic Aspects

- edema prophylaxis:(1) elevation and compression, (2) isometric and dynamic exercises, (3) possibly ice applications
- circulation stimulation:(1) isometric and dynamic exercises, (2) ice applications, (3) other modalities, (4) foot baths
- mobilization: (1) active movements, (2) foot baths
- strengthening: (1) active exercises, (2) resistive exercises
- gait training

Course of Treatment

In the first six weeks, during the immobilization period, the therapist should work with the patient, following previously described guidelines for edema prophylaxis. From the seventh week on, following removal of the cast, the rehabilitation program starts with foot baths and dynamic and resistive exercises. The objective is to achieve free motion of the tarsal joints through strengthening of the foot and leg muscles. The patient is allowed to begin weightbearing after being supplied with a compression stocking and an arch support. In most cases, no additional support and only a short period for gait correction are required.

Sample Exercises

In the final phase of gait training, with the patient standing between parallel bars or without support, the following exercises may be performed:

- swinging the affected leg forward and backward, side to side, in a circle
- shifting the weight when stepping forward, backward, and side to side
- practicing selected steps, forward, backward, and to both sides
- hopping
- standing on the toes, singly and bilaterally
- walking on tiptoes
- rhythmic lower extremity exercises to music, dancing

FRACTURES OF THE METATARSAL BONES

Mechanism of Injury

Fractures of the metatarsal bones are usually caused by a direct impact.

Types

The fractures are described as fractures of the head, shaft, and base.

Clinical Considerations and Diagnosis

Edema, bony displacement, and pain are produced. The diagnosis is confirmed by radiographic examination.

Medical Management

The treatment is directed at preserving and restoring the arches of the foot. Early realignment of the heads of the metatarsals is essential to permit later weightbearing. Usually, surgery is required for fractures of the first and fifth metatarsals and for dislocated fractures of the bony tuberosity of the fifth metatarsal. Wires, screws, and plates are used for internal fixation.

Complications

Uneven intra-articular surfaces can cause problems in walking and must be corrected by an osteotomy.

Fractures of the First Metatarsal Bone

Symptoms and Therapeutic Aspects

The symptoms and therapeutic aspects are the same as for nonsurgically treated fractures of the tarsal bones.

Course of Treatment

For the first six weeks, in both surgical and nonsurgical treatment, the patient is put in a cast and is instructed in edema prophylaxis. In the seventh week, as soon as the cast is removed, the rehabilitation program begins with hydrotherapy, dynamic exercises to mobilize the toes and proximal and distal tarsal joints, and exercises to strengthen the muscles of the foot and leg.

The patient is given a compression stocking and arch support and is allowed to walk with full weightbearing, initially with the aid of crutches or canes. Training and correction of the gait may also be necessary.

Sample Exercises

In the final phase of mobilization, strengthening, and gait training, the exercises described earlier for fractures of the calcaneus and for nonsurgically treated fractures of the tarsal bones should be performed.

Nonsurgically Treated Fractures of the Second Through Fifth Metatarsals

Symptoms

With the foot end of the bed slightly elevated, and in some cases, with the injured leg placed in a splint, the following symptoms are observable:

- edema and discoloration around the fracture site
- limited range of motion, especially of the toe joints, due to pain and swelling
- pain around the fracture site

Therapeutic Aspects

The therapeutic aspects are the same as those for nonsurgically treated fractures of the small tarsal bones.

Course of Treatment

Beginning on the first day, the patient is given cold foot baths and is instructed in isometric and dynamic exercises. The patient exercises all movements, without weightbearing, to regain the full range of motion of all joints and to maintain strength of all muscles of the foot. In addition, the patient receives a compression stocking and is permitted to walk with crutches and without weightbearing.

After about the seventh week, the patient can begin full weightbearing, but with an arch support. Support from crutches or canes is withdrawn as soon as possible. If there are no major problems, the patient should participate in foot and walking exercises.

Sample Exercises

In the middle and late phases of mobilization, strengthening, and gait training, the patient should perform the exercises previously described for surgically treated fractures of the tibial shaft, surgically treated fractures of the proximal tarsal joint, and nonsurgically treated fractures of the small tarsal bones.

FRACTURES OF THE PHALANGES

Mechanism of Injury

Toe fractures are caused mainly by direct trauma. In a considerable number of cases, they are compound fractures, for example, caused by accidents with lawnmowers.

Types

Fractures of the toes are classified as fractures of the shaft or as fractures occurring close to the joint. The types of fractures are transverse, oblique, or comminuted.

Medical Management

A slight dislocation or fracture is treated with an adhesive bandage. A more severe fracture or dislocation, especially of the first metatarsal, can be treated surgically. Since the big toe is very important for pushoff during gait and balancing on the toes, a prolonged period of immobilization and nonweightbearing is indicated, with a lower leg cast for six weeks. The other toes are not essential for standing and walking; in such cases, a certain degree of dislocation can be tolerated, and treatment with an adhesive bandage is sufficient. In open fractures of the fifth toe, a splint is indicated because of the considerable soft-tissue swelling.

Early and Late Complications

Continued joint stiffness can cause gait deviations and calluses. Occasionally, toes with open fractures must be amputated.

Nonsurgically Treated Fractures of the Toes

Symptoms

The symptoms for nonsurgically treated fractures of the toes are the same as for nonsurgically treated fractures of the second through fifth metatarsal bone.

Therapeutic Aspects

The therapeutic aspects are the same as for nonsurgically treated fractures of the tarsal bones.

Course of Treatment

Beginning on the first day, the resorption of the hematoma and swelling is accelerated by the application of ice, and dynamic exercises of the leg and foot are begun. In the first few days, the patient might need crutches or a cane for support during walking. Later an arch support may be needed.

Sample Exercises

In the middle and late phases of mobilization, strengthening, and gait training, the patient should perform the exercises previously described for surgically treated fractures of the proximal tarsal joint, fractures of the calcaneus, and nonsurgically treated fractures of the small tarsal bones.

Chapter 11

Injuries of the Shoulder Girdle

The shoulder girdle connects the arm to the thorax. The clavicle, the scapula, the upper arm, and the wall of the thorax form a functional unit. The individual elements are connected by joints and ligaments. A syndesmosis connects only the clavicle and the sternum; the remaining shoulder girdle is joined to the thorax exclusively by muscles. This provides for maximum mobility of the upper arm and the portions of the shoulder girdle that form the shoulder joint.

The anatomical structure of the shoulder complex allows for a variety of injuries. The clavicular fracture is one of the most frequent of fractures, and traumatic subluxations of the shoulder joint occur more often (50 percent of all cases) than any other type of subluxation. Lesions in one area of the shoulder girdle can interfere with the function of the entire functional unit.

CLAVICULAR FRACTURES

Among the fractures of the bones that form the shoulder girdle (scapula and clavicle), clavicular fractures occur most frequently.

Mechanism of Injury

The main causes are an indirect force, resulting from a fall on the extended arm or shoulder, a direct impact, or, sometimes, extreme muscle contraction.

Types

Oblique and comminuted fractures affect mainly the medial third. The acromial third is less frequently affected. The sternal third is rarely affected.

Symptoms, Clinical Considerations, and Diagnosis

The symptoms are swelling and deformation, crepitation, and pain during movements of the shoulder girdle. The diagnosis is confirmed by radiographs in anterior, posterior, and oblique planes.

Medical Management

The treatment is usually nonsurgical. A Desault bandage is applied for 3 to 4 weeks, and exercises are started early. A plate osteosynthesis is indicated in more severe dislocations, in cases where there is danger of skin penetration, and in cases with accompanying lesions of the plexus and blood vessels.

Complications

Neurological deficits caused by hypertrophic callus formation require that the hypertrophic bone be resected. Painful pseudarthroses have to be stabilized surgically by plates.

Physical Therapy

The following therapeutic program is applicable to nonsurgically treated clavicular fracture.

Symptoms

The patient is usually treated with a Desault bandage. The following symptoms are observable:

- swelling and hematoma across the fracture site
- crepitation of the fracture
- limited mobility of the arm on the injured side due to pain
- pain

Therapeutic Aspects

- mobilization: (1) relaxation techniques with ice, (2) dynamic muscle work, with and without equipment, (3) joint mobilization
- strengthening: (1) dynamic exercises against resistance and with equipment, (2) resistive exercises
- exercising and training in daily skills

Course of Treatment

Clavicular fractures do not require bed rest and often not even physical therapy. If therapy is prescribed, it is tailored to individual needs. Usually, the purpose of such therapy is to maintain the mobility of the shoulder joint in all directions or, if mobility has been lost, to restore it.

Sample Exercises

The following exercises are used in the middle and late phases of mobilization, strengthening, and training in daily activities. In the first set, PNF exercises are done in a supine position, while sitting, and in a prone position with chest and arms unsupported (projecting over an edge):

- with extended elbow exercising the diagonal from an extension-adduction-internal rotation to a flexion-abduction-external rotation; if the injury is minor, attempt to resist at the end position of the diagonal. Gentle stretching is permissible with full approximation avoided at first

As a second set of exercises, the patient can do those listed as a second set for fractured single ribs.

SCAPULAR FRACTURE

Mechanism of Injury

The scapular fracture can be caused by either a direct or an indirect force such as through a fall where force is transmitted through the humerus.

Types

Scapular fractures are rare but they can occur in the neck, the acromion, the coracoid, or the body of the scapula (see Figure 11–1).

Clinical Considerations and Diagnosis

The symptoms of scapular fractures are pain around the fracture area and on palpation and a limited ability to raise the arm. Radiographs are used to confirm the location of the fracture.

Medical Management

The treatment of a scapular fracture is usually nonsurgical. Depending on the extent of the fracture, it is immobilized for one week in a Desault bandage or for

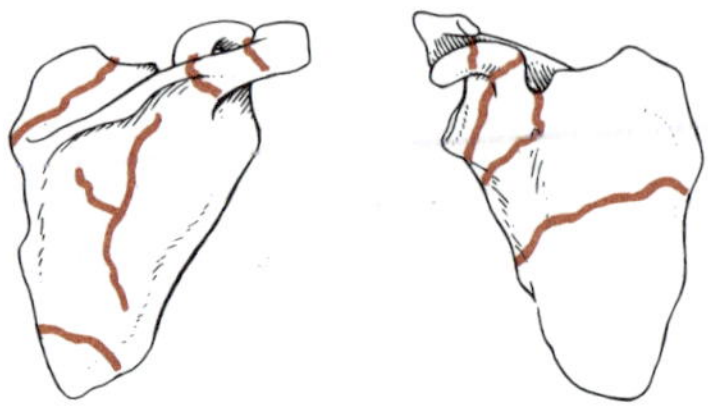

Figure 11–1 Fractures of the scapula: fracture of the scapular ridge, coracoid fracture, acromion fracture, neck of the scapula, fracture of the body of the scapula.

Source: Schlosser and Kuner, 1980

two weeks in an abduction airplane splint. This is followed by physical therapy. The Desault bandage is recommended if there are no, or only a few, displaced fractures of the scapula and the fragments are tightly fixed. In more severely displaced fractures, an adduction position must be avoided to prevent a contracture of the shoulder joint. Surgical reposition and fixation is required for fractures with grossly displaced fragments, especially if they are interfering with the congruency of the shoulder joint. Early exercise treatment is mandatory. The main concern is the prevention of a frozen shoulder.

Early and Late Complications

The subscapular nerve may be involved, with a loss of external rotation strength, as well as the suprascapular nerve. Frozen shoulders need intensive physical therapy and may also need mobilization under general anesthesia.

Physical Therapy

The following therapeutic program is applicable to nonsurgically treated fractures of the scapula.

Symptoms

With the arm of the injured side resting on an abduction pillow, the following symptoms are observable:

- possibly hematoma and swelling in the area of the scapula
- painfully limited range of motion of shoulder and arm of the injured side
- possibly painful thoracic expansion
- general pain

Therapeutic Aspects

- pneumonia prophylaxis: (1) breathing exercises
- support for local blood flow: (1) heat or cold, (2) isometric and dynamic exercises
- mobilization: (1) relaxation techniques, possibly with ice, (2) dynamic exercises, with and without equipment, (3) scapula mobilization
- strengthening: (1) dynamic exercises against resistance with equipment
- training in daily activities

Course of Treatment

Usually, the patient is permitted to get up immediately and exercise in all positions. Ultrasound may be helpful in improving the resorption of considerable contusions and hematomas.

In the first few days, breathing exercises are necessary in most cases. The therapist tries to counteract the painful restrictions of the thoracic expansion and the risk of pneumonia by combining active respiration and thorax-enlarging exercises.

At the same time, the extremity of the injured side is strengthened and mobilized. Initially, most patients have difficulty exercising in a supine position because of pressure pain. In such cases, exercises in a side position or while sitting and standing are acceptable alternatives.

Bilateral exercises in front of a mirror can supplement these exercises. The patient can also work with a cane or ball in all exercise positions while lying in a prone position on a platform mat (arms over the edge). The rehabilitation program may conclude with a home program and swimming exercises.

Sample Exercises

The following exercises are used in the middle and late phases of mobilization and strengthening. In the first set, the patient lies in a prone position on a platform mat, with the injured side close to the edge and using a gymnastic ball:

- rolling the ball on the floor toward the head (flexion), then toward the feet (extension)
- rolling the ball to the side, then underneath the mat
- rolling the ball from underneath the mat (extension-adduction) diagonally upward (flexion-abduction)
- rolling the ball in a wide semicircle from the feet over the side to the head and back

If this exercise is too difficult for the patient, it can be modified by using a larger ball (less stretching, more support).

As a second set of exercises, the patient may repeat all of the above with a rod—pushing it, touching the floor with it, and lifting it up at the end of each moving phase. The rotary components (PNF rotations) of these exercises can in fact be performed better with a rod than with a ball.

A home exercise program may involve clinical sports, such as games of darts, badminton, or table tennis.

INJURIES OF THE ACROMIOCLAVICULAR JOINT

Among the joints of the shoulder girdle, the acromioclavicular joint has the greatest risk of injury.

Mechanism of Injury

The primary cause of an injury to the acromioclavicular joint is a fall on the shoulder.

Types

In addition to a subluxation of the shoulder joint, the capsule and acromioclavicular ligaments may be ruptured (see Figure 11–2).

In the classification by Tossy, there are three types of injury to the acromioclavicular joint:

1. stretching with partial rupture of ligaments (acromioclavicular and coracohumeral), with the lateral end of the clavicle not elevated significantly
2. rupture of the acromioclavicular ligament and the joint capsule and partial rupture of the coracoclavicular ligament, with the clavicle elevated peripherally one half the width of the shaft
3. complete rupture of all ligaments, with the clavicle elevated up to the full width of the shaft

Clinical Considerations and Diagnosis

The disruption of the acromioclavicular joint is characterized by a ridge in the shoulder line. The end of the clavicle is elevated, and, on pressure, it springs back elastically (keyboard phenomenon). The dislocation can be verified radiographically with the arm being pulled down by holding 10 kg weights.

Medical Management

Tossy Type 1 injuries are treated nonsurgically. An adhesive bandage can be applied temporarily. After the pain has subsided, physical therapy is started. The

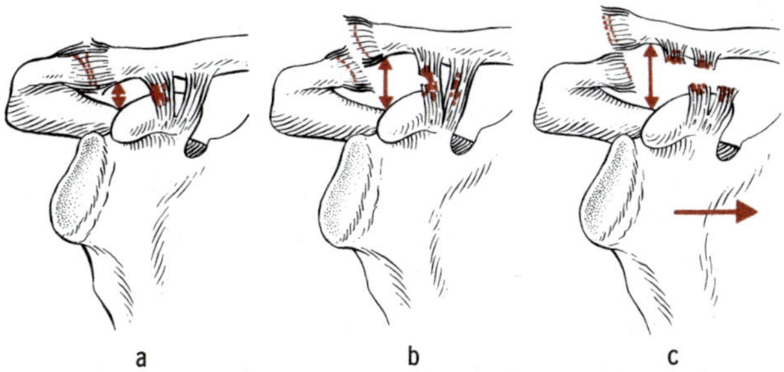

Figure 11–2 Degrees of severity of shoulder girdle dislocation as classified by Tossy: a. internal strain; b. partial tear; c. complete tear of the ligamentous attachment.

Source: Schlosser and Kuner, 1980

treatment of Tossy Type 2 injuries is usually also nonsurgical; an airplane splint is used with a padded abduction support. Surgery is usually indicated for Tossy Type 3 injuries. In such cases, the capsule and ligaments are repaired with sutures and temporary (8 to 12 weeks) internal fixation. For the latter procedure, Bosworth suggests a wire fixation for the acromioclavicular joint or a screw fixation of the clavicle onto the coracoid.

Late Complications

In chronic subluxations of the acromioclavicular joint, a ligament reconstruction might be indicated.

The following therapeutic program is applicable to a surgically treated disruption of the acromioclavicular joint (ligament suture and fixation by a screw).

Physical Therapy

Symptoms

After removal of the fixating device and later of the airplane splint, the following symptoms are observable:

- pitting edema
- atrophy around the deltoid muscle, often an inability to reach a fair grade in a muscle test
- limited range of motion of shoulder joint
- slight pain during movements above and below the level of the immobilization position

Therapeutic Aspects

- pneumonia prophylaxis: (1) breathing exercises, (2) active movement, (3) changing positions
- edema prophylaxis: (1) position (abduction splint, abduction wedge pillow), (2) compression bandage, (3) intermittent pressure and effleurage massage, (4) ice applications, (5) isometric and dynamic exercise
- mobilization: (1) position (abduction splint, abduction wedge pillow), (2) relaxation techniques with ice, (3) active movement with and without equipment, (4) joint mobilization
- strengthening: (1) dynamic exercise against resistance, (2) PNF, (3) therapeutic pool
- training in activities of daily living: (1) at first, changing positions, getting up, and engaging in necessary daily activities with the airplane splint; (2) after removal of the splint, exercising daily skills

Course of Treatment

Beginning on the first day, because the airplane splint compresses the chest, it is frequently necessary to teach the patient breathing exercises. Most patients have to learn how to live with the splint, that is, changing positions, getting up, and engaging in daily activities.

While in the splint, the patient can begin to do dynamic exercises with the fingers, wrist, and elbow joint and isometric contractions of the deltoid muscle.

If Kirschner wires have been implanted, they have to be removed prior to physical therapy, since they would break during the exercises. The screw fixing the clavicle to the coracoid process is not affected by these movements.

For a short time, an abduction splint supports the weightbearing of the arm (see Figure 11–3). During the nonweightbearing phase, the arm is placed on an abduction pillow. Both devices have a prophylactic effect on the edema; they also prevent overstretching of the deltoid muscle and the capsular ligaments of the shoulder. Edema prophylaxis is enhanced by gentle effleurage massage in an elevated position; by intensive isometric muscle contractions, in combination with ice applications; and by dynamic exercise. Sometimes the swelling is so pronounced that the arm has to be wrapped in an elastic bandage.

The full range of motion of the distal joints (finger, hand, elbow) is quickly regained; however, the elimination of the contracture in the shoulder joint requires more time and effort. In addition to the use of relaxation techniques with ice and active movement, it is recommended that the patient work with equipment such as a ball or cane in different positions. Bilateral exercises in front of a mirror are beneficial. Simultaneously with mobilization, strengthening exercises (PNF) against controlled resistance and active work with equipment complete the

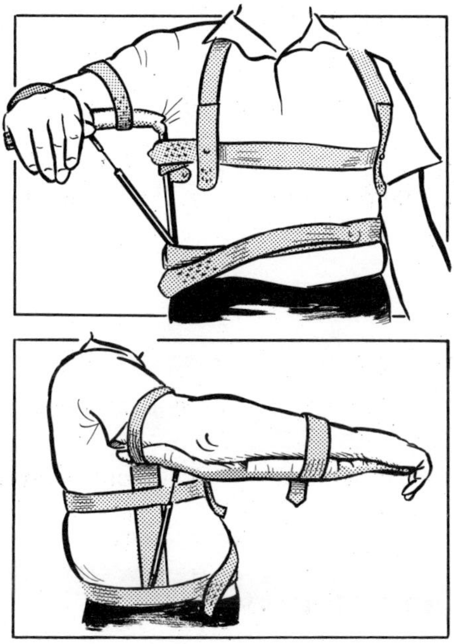

Figure 11–3 Abduction splint

rehabilitation program. Under no circumstances should the extension be forced at the beginning of treatment over the neutral position or against resistance, with the risk of straining ligaments.

During the following weeks, the treatment includes swimming, increased work against resistance, exercises with equipment, and training in activities of daily living by occupational therapy.

Sample Exercises

The following exercises are used in the early phase of supporting local blood flow, mobilization, and strengthening. While doing these exercises, the patient is in a bivalved abduction cast.

1. finger
    - making a fist, spreading the fingers
    - pressing a sponge
    - thumb-finger opposition exercises
    - placing in position opposite the thumb
2. wrist: flexion, extension, adduction and abduction, circling the hands in both directions

3. elbow, with the arm supported, one hand above the olecranon, the other one near the wrist under the forearm

   - flexion, extension
   - supination, pronation
   - flexion-supination, extension-pronation, always including wrist and fingers in the exercise

4. shoulder

   - isometric contractions of the deltoid muscle
   - dynamic exercises, elevating the arm out of the cast and holding it without support
   - with the shoulder in an elevated position, moving it slightly in horizonal adduction/abduction

INJURIES OF THE STERNOCLAVICULAR JOINT

Mechanism of Injury

Injuries of the sternoclavicular joint are caused mainly either by a direct impact on the clavicle or indirectly by a contusion, possibly in conjunction with an injury of the sternum.

Types

The dislocation can occur in an anterior or posterior direction. A posterior dislocation of the clavicle is dangerous when it affects the mediastinum.

Clinical Considerations and Diagnosis

Pain and dislocation are characteristic of an injury of the sternoclavicular joint.

Medical Management

A subluxation is treated nonsurgically. It is reduced by pulling the shoulders outward and to the back. Afterward, a Desault bandage is applied. If the subluxation cannot be corrected, the joint has to be reduced surgically and fixed temporarily with a wire loop. Immobilization in an airplane splint is required.

Early and Late Complications

Simultaneous injuries of adjacent structures can occur. Also, an incomplete reposition can cause complaints, especially in one who does manual labor. The medial end of the clavicle is stabilized by ligaments.

INJURIES OF THE SHOULDER JOINT

Following a shoulder injury with associated pain and disturbed function, a diagnosis of a shoulder sprain is indicated only if no ruptures of the ligaments or bone lesions appear radiographically. Injuries from severe impacts that do not show radiographically recognizable "micro fractures" can occasionally result in partial necroses of the joint surfaces. The treatment of a shoulder contusion is therefore dependent on the circumstances of the accident, the clinical examination, and the treatment's progress.

Mechanism of Injury

Injuries of the shoulder joint are caused by forces that force the upper arm beyond its normal range of motion or direct impacts from falling on the shoulder or arm.

Types

Contusions and sprains can cause persistent pain and limited range of motion, especially in the area of the shoulder and the cervical spine.

Clinical Considerations and Diagnosis

There are no external symptoms. Pain can be caused by palpating the greater tuberosity; also, the range of motion is painfully limited, especially during external rotation and elevation.

Medical Management

After initial treatment with a Desault bandage (3 to 4 days) and reduction of the acute pain, an intensive physical therapy is started, supported by analgesics and anti-inflammatory medication. In case of persistent complaints or an avulsion fracture, the joint has to be placed in abduction and exercises must be started from this position.

In the case of an injury leading to a rupture in a patient up to 40 years of age, a suture repair of the tendon should be considered if there is clinical evidence of a ruptured rotator cuff. In older patients with ruptures due to minor trauma, the treatment is nonsurgical.

Complications

The main concern is a stiff shoulder. This requires intensive physical therapy and, sometimes, mobilization under general anesthesia.

TRAUMATIC SUBLUXATION OF THE SHOULDER JOINT

Mechanism of Injury

A subluxation of the shoulder is caused by a strong direct or indirect force, such as from falling on the shoulder or from an abrupt movement of the arm. If there is a preexisting dysplasia, the subluxation might occur after a minor trauma.

Types

A subluxation of the shoulder joint is common because of the anatomical structure involved: a flat socket with insufficient bony containment of the humeral head. The head is held in place only by its joint capsule, the long head of the biceps tendon, the coracohumeral ligament, and the tendons of the rotator cuff. Given this anatomical structure, the subluxation occurs because of a forced external rotation of the humerus, forcing the head of the humerus forward and downward,

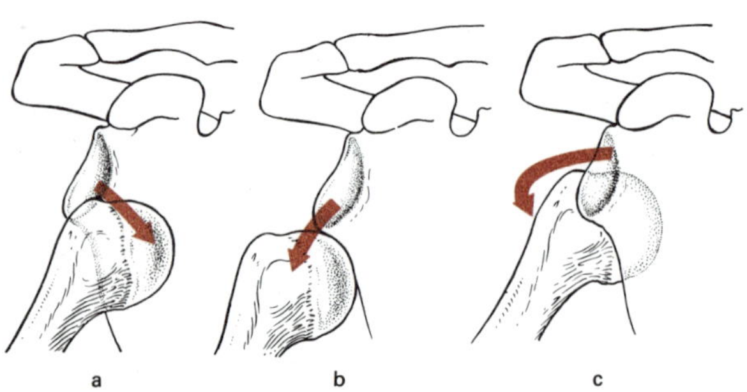

Figure 11–4 Types of shoulder subluxations: a. in the anterior direction and downward; b. in the lower subglenoid direction; c. in the posterior direction and downward.

Source: Schlosser and Kuner, 1980

which is occasionally combined with a rupture of the glenoid labium and the joint capsule. Posterior subluxations are observed less frequently (see Figure 11–4). Frequently, the subluxation of the shoulder occurs with a bone lesion, such as a separation fracture of the greater tuberosity or, less frequently, with a fracture of the humeral head.

Clinical Considerations and Diagnosis

A subluxated shoulder is characterized by a change in shoulder line. In an axillary subluxation, the head is displaced downward into the armpit. The deltoid muscle loses its typical round shape. The socket is empty, and the shoulder appears to be flattened. In a subcoracoid subluxation with dislocation of the humeral head across the anterior rim of the socket, the shoulder contours appear sharp. Motion of the arm within the shoulder joint is lost. In a posterior subluxation, the socket is empty and the head can be palpated dorsally. Passive mobilization attempts can create spontaneous pains. A radiograph can confirm the diagnosis and reveal simultaneous bone lesions. A bone lesion in the shoulder joint without a subluxation has to be excluded as a differential diagnosis. Such a lesion can produce similar symptoms but without the loss of range of motion.

Medical Management

The treatment of choice is an immediate repositioning of the humeral head under general anesthesia. The arm is pulled into slight abduction and external rotation. The humerus is then slowly internally rotated and horizontally adducted across the chest, achieving reduction of the subluxation. In an accompanying subcapitular fracture, the procedure can be nonsurgical, controlled by an image intensifier. In a pending separation of the fragment, an open reduction is required. Surgery is also necessary for a chronic subluxation of the shoulder. After a short period of immobilization (3 to 4 days), active physical therapy is started.

Complications

Circulatory disturbances in the humeral head, especially nerve deficits, are possible. During a fracture or subluxation and also during a severe distortion whose basic circumstances are similar to those of a subluxation, lesions of the brachial plexus or partial ruptures of nerve roots or portions of the brachial plexus can occur. The axillary nerve is frequently involved. Thus, a variety of lesions—even a complete paralysis of the brachial plexus—can result from a traumatic subluxation of the shoulder joint.

Physical Therapy

The following therapeutic program is applicable in nonsurgical treatment of younger patients.

Symptoms

The shoulder is put in a Desault bandage for up to four days. The following symptoms are observable:

- possibly hematoma, swelling around the shoulder
- sometimes swelling of the entire arm
- limited range of motion of the joint
- pain

Therapeutic Aspects

- edema prophylaxis: (1) positioning on an abduction pillow, (2) compression bandage, (3) isometric and dynamic exercise, (4) ice applications
- strengthening: (1) isometric exercises, (2) dynamic exercises against resistance, (3) electrical stimulation
- mobilization: (1) positioning, (2) relaxation techniques with ice, (3) active movement with and without equipment, (4) therapeutic pool, (5) joint mobilization
- training in activities of daily living

Course of Treatment

During the first four days, the patient exercises with the normal arm (dumbbells, pulleys, and so on) or with the therapist (PNF). The patient also works with the fingers, wrist, and elbow on the injured side. Beginning on the fifth day, to provide muscular support for the shoulder joint, the treatment emphasizes strengthening more than mobilization. The patient initially exercises the arm in a supine position; subsequently, while sitting in front of the mirror, the patient exercises the arm with isometric and dynamic exercise. In these exercises, movement in the direction of the subluxation must be avoided. Also, for about three weeks, flexion or abduction in the shoulder joint of more than 90 degrees should not be included in the exercises. Flexion over 90 degrees would stretch the joint capsule, and abduction over 90 degrees would combine with external rotation and stress the shoulder joint. In any event, extension or external rotation past the neutral position is not permitted during the first weeks. A severe hematoma can be resorbed by ultrasound. If the deltoid muscle is very weak, the therapy should include the use of electrical stimulation. Beginning in the third week, the patient can exercise in the therapeutic pool. To supplement the strengthening and mobilization, training in daily activities should be started carefully. After about seven weeks, the exercises in the therapeutic pool are replaced by swimming, and occupational therapy is started for activities of daily living.

The therapeutic program for older patients suffering from traumatic subluxation of the shoulder joint is basically the same as that for young patients. However, since the risk of a post-traumatically stiff shoulder is greater for older patients, their exercises should begin on the first postoperative day, in contrast to starting on the fifth postoperative day for younger patients.

Sample Exercises

The following exercises are used in the early and middle phases of strengthening and mobilization. In the first set, the patient is in a supine position, with the arm in slight flexion-abduction and placed on an abduction pillow.

- isometric exercises of all arm muscles
- elevating the arm slightly and holding it without support

The second set of exercises is done while the patient is sitting, with the arm on a gymnastic ball and the shoulder in approximately 40 to 50 degrees of flexion and 30 to 40 degrees of abduction.

- holding this position against carefully increased resistance
- rolling the ball, with the arm, short distances forward and back again, without going into extension

While the therapist holds the ball, the patient tries to elevate the arm from the initial position on the ball, and to hold it up without support for a short period of time.

In the third set of exercises, the patient sits on a stool in front of a mirror and moves a cane with both hands within a 90 degree arc for flexion and abduction of the shoulder. The patient should use the shoulder muscles to achieve flexion and abduction and not substitute with trunk motion (i.e., arching the back) to achieve 90 degrees of flexion.

RECURRENT SUBLUXATION OF THE SHOULDER JOINT

Mechanism of Injury

Recurrent shoulder subluxation develops from a previous severe distortion or traumatic subluxation when the ruptured capsular ligaments have not healed sufficiently. In such situations, it can occur without an external trauma. The recurrent subluxation may be triggered by normal movements and can be reduced easily.

Types

The recurrent subluxation takes the same forms as the original subluxation.

Medical Management

After the third subluxation, the treatment is surgical. Two different techniques have proved successful: the chronic subluxation is treated either by a rotation osteotomy immediately below the humeral head or by elevating the rim of the socket (implanting a bone chip). In either case, the procedure is combined with a shortening of the tendon of the subscapular muscle and a tightening of the joint capsule. Immediately after the osteotomy with stable osteosynthesis, physical therapy can be started. The surgical elevation of the rim of the socket and soft-tissue tightening has to be immobilized for six weeks. This is followed by active exercises.

Complications

Resubluxation is possible.

INJURIES OF THE BRACHIAL PLEXUS OR THE AXILLARY NERVE

If the trauma symptoms include neurologic damage in the form of lesions of the brachial plexus or axillary nerve, the normal treatment plan must be expanded. In this case, to the extent the basic injury permits, the following procedures are added to the therapy:

- support by an abduction pillow in bed or an abduction splint outside the bed
- support of local blood flow with exercises in the therapeutic pool
- electrotherapy
- short, quick ice applications, in combination with exercises for weak muscles
- stimulating and strengthening exercises in the therapeutic pool as soon as the basic injury allows it, utilizing the buoyance of the water and buoys placed in loops to support the weight
- PNF exercises
- training in daily activities

Patients with lesions of the brachial plexus or axillary nerve should also participate in occupational therapy for training in activities of daily living.

Chapter 12

Injuries to the Humerus

Fractures of the humerus are classified as proximal fractures (fractures of the humeral head and neck and tuberosity), as fractures of the shaft, and as distal humeral fractures. Proximal fractures are frequently associated with injuries of the shoulder joint. Distal fractures can be associated with injuries of the elbow joint.

HUMERAL HEAD AND NECK FRACTURES

Mechanism of Injury

Fractures of the humeral head and neck are usually caused by a direct impact, but they can also be caused indirectly by falling on the shoulder, elbow, or hand.

Types

Fractures of the humeral head and neck are classified as subcapital humeral fractures, separation fractures of the greater tuberosity, and subluxation fractures (see Figure 12–1).

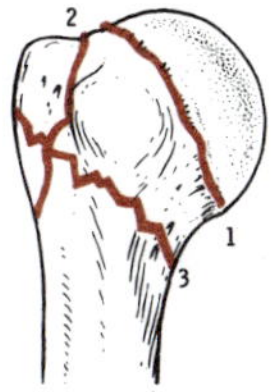

Figure 12–1 Types of fractures of the humeral head: 1. fracture in anatomical shaft; 2. tear of the tuberculum major; 3. fracture in the surgical shaft of the humerus.

Clinical Considerations and Diagnosis

The symptoms frequently resemble those for a subluxated shoulder in many cases. However, instead of a loss of motion, a painfully limited range of motion and crepitation result. In a subluxation fracture, the socket is empty. Radiographic examinations are used to confirm the diagnosis.

Medical Management

The treatment of a proximal humeral fracture is usually functional. Generally, a good functional result is more important than a radiographically perfect alignment of the fracture. This is explained by the fact that the upper extremity—used mainly to grasp or carry—is less exposed to axial loads than the leg. Thus, malpositions do not result in abnormal weightbearing of the neighboring joints. This allows early mobilization to provide a good functional restoration.

Frequently, fixed fractures of the anatomical neck do not require any immobilization. If the fragments are slightly displaced, a Desault bandage is applied for a short period but without repositioning. The exercises, especially for older patients, should be started no later than five or six days after the above treatment, even if the treatment included repositioning. Bandages to fix the upper arm to the chest should be applied only temporarily, since the prevention of a contracture has priority, and in fact is mandatory for older patients.

Frequent fractures at the surgical neck with moderate adduction, abduction, and/or sagittal deviation are treated with a hanging cast. This permits the fracture to heal in a comfortable position without limiting excessively the patient's mobility.

Surgery is indicated in severely dislocated fractures, comminuted fractures, dislocated separation fractures of the greater tuberosity, and subluxation fractures that cannot be reduced. The fragments are stabilized with screws or buttress plates. Postoperatively, after a short period of immobilization, exercises are started with the patient in an abduction splint.

SUBCAPITAL FRACTURE OF THE HUMERUS

Medical Management

Basically, the treatment is nonsurgical and early functional, as modified by Poelchen. Patients with reduced or nonreduced fractures are usually placed in the following position: A dorsal shoulder-arm-hand cast is applied and adhesive tape is used extending from the shoulder to the hand. A weight is hooked to the distal end. The patient's arm is placed on a special abduction bench in at least 30 degree abduction. Based on a radiological examination, a 200 to 400 g weight is situated to pull at the end of the bandage (see Figure 12–2).

On the day of the accident, the case is bivalved and for the next two weeks is bandaged on only during the night. When the patient rests in bed, the weight traction should be applied; when the patient stands up and moves around, the weight of the arm itself replaces the traction.

In modern trauma care, the patient no longer carries the weights around, since that increases the risk of overstretching the muscles and subluxating the shoulder joint. The patient is encouraged to press and knead a foam ball as often as possible; this is beneficial for peripheral muscle tone, and increasing the blood flow.

While in a sitting position, the patient should keep the injured arm on the thigh, on an armrest, or on a table. After about two weeks, the abduction bench is replaced by an abduction pillow.

Symptoms

- swelling of the distal areas, especially hand and finger
- pain during rest, increasing when changing positions and moving
- sometimes reduced circulation and respiration

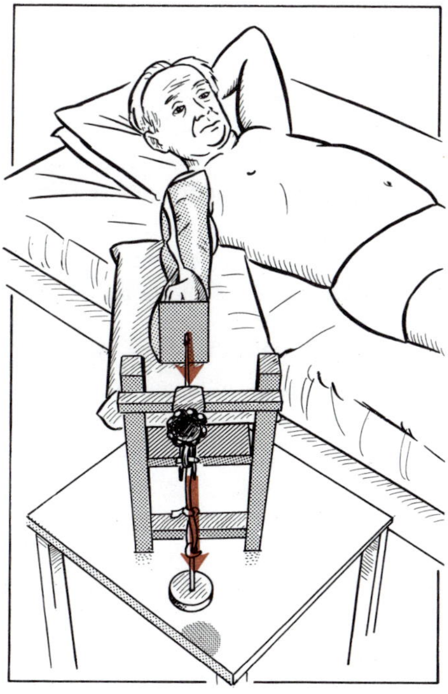

Figure 12–2 Positioning of the arm on an abduction bench

Therapeutic Aspects

- edema prophylaxis: (1) kneading a foam ball, (2) positioning, (3) compression bandage, (4) ice applications, (5) isometric and dynamic exercises
- mobilization: (1) positioning, possibly with abduction splint, (2) active movements, PNF, (3) relaxation techniques with ice, (4) joint mobilization
- strengthening: (1) active movements, (2) exercises against resistance (PNF), (3) electrical stimulation
- training in activities of daily living

Course of Treatment

On the first day, the therapist tries to encourage the patient to cooperate, giving the patient relevant information to diminish the anxiety and insecurity. Physical therapy starts with exercises of the fingers and finger manipulation of the foam ball.

On the second day, the cast is removed. The shoulder joint remains immobilized, and the fingers, wrist, and elbow are moved actively. The shoulder muscles, especially the deltoid muscle, should be contracted isometrically.

After the third day, if the tone of the deltoid muscle is satisfactory, the patient can get up. The first few times, the patient will need support in doing this. In the case of a large hematoma, modalities can support the resorption and reduce discomfort.

As shown in Figure 12–3, the patient bends forward and swings the arm in a sagittal plane, pressing the foam ball tightly in the hand. If the patient has difficulty letting the arm go into the swinging motion and tries to move it actively instead, the patient should hold a little weight in the hand, thus making the hand feel like a pendulum in a passive swinging motion. The patient should exercise independently as often as possible. On the fifth day, at the latest, swinging motions in a frontal plane are added to the program.

After the seventh day, the arm can be swung in a circle in both directions. The therapist must keep encouraging the patient to exercise independently. While exercising with the trunk in a forward position, increasingly active movements and holding phases are added between the swinging motions. A very weak deltoid muscle may need additional electrotherapy.

Beginning in the second week, mobilization and strengthening can be done in supine, sitting, and standing positions. Exercises in rotation should be done very carefully and only insofar as necessary for retraining in daily activities. At no time should the patient's pain threshold be exceeded.

Beginning in the third week, the exercises are increased in frequency and difficulty. Internal and external rotation movements are now allowed. Patients with stable circulation should exercise preferably in a prone position at the edge of

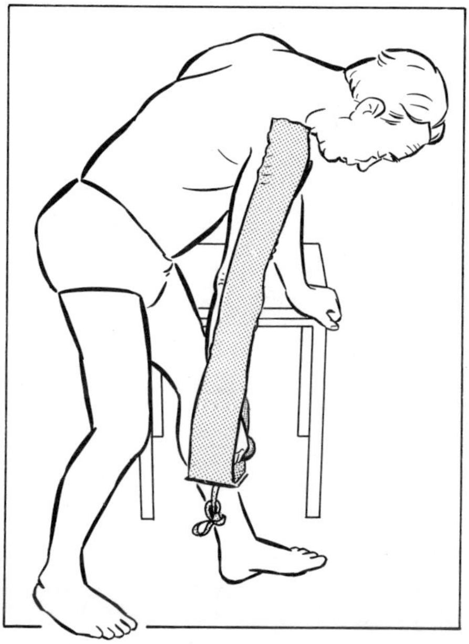

Figure 12–3 Starting position for pendulum exercises

a platform mat, using a cane, gymnastics ball, or other kind of ball to increase range of motion. In all these exercises, the use of equipment should gradually be increased. The patient will be able to develop a better awareness of body movements, and to correct them as necessary, if the exercises are done in front of a mirror. Exercises in the therapeutic pool can facilitate the treatment activities; the buoyancy of the water supports a still limited range of motion and most patients enjoy exercising in the pool.

Beginning in the fourth week, the therapy is supplemented by swimming and occupational therapy. Strong resistance must not be used until there is a complete consolidation of the fracture.

Sample Exercises

The following exercises are used in the middle phase of mobilization and strengthening in the therapeutic pool. The first set is done with the patient sitting or standing up:

- bringing a ball with both hands toward the body, then holding it away from the body again with extended arms

- pulling the extended arm from extension alongside the body into shoulder flexion and external rotation to the surface of the water, then bringing it back with internal rotation
- doing the same exercise with shoulder abduction and external rotation, then bringing the arm back with internal rotation
- at a level just below the surface of the water, bringing the hand of the injured side from the opposite shoulder with extended elbow and supination into shoulder abduction and external rotation, then back again
- in a PNF exercise, moving the arm from extension-abduction-internal rotation deeply downward into flexion-adduction-external rotation, with both flexed elbow and extended elbow
- doing the same exercise with the other diagonal

In the second set of exercises, the patient is bent forward, while standing, with the normal arm on the pool edge as a support. The arm is brought up from the water with extended elbow into shoulder flexion and external rotation to the water surface. Then the same exercise is done with a shoulder abduction and external rotation.

In the third set of exercises, the patient is in a prone position at the edge of the pool, with the injured arm hanging down in the water. The patient moves the extended arm into shoulder flexion and external rotation, first to the surface of the water, then above the surface. The patient may do the exercise using underarm support in a therapeutic pool as described earlier for a clavicular fracture.

Finally, the patient may do the exercise in a supine position in a therapeutic pool as described earlier for a clavicular fracture.

HUMERAL SHAFT FRACTURES

Mechanism of Injury

Fractures of the humeral shaft are caused by a direct impact on the upper arm or, more frequently, by an indirect force, for example, a fall on the elbow or hand (see Figure 12–4).

Types

A direct impact usually results in uncomplicated transverse fractures or comminuted fractures. Indirect forces cause spiral fractures and comminuted fractures with wedge-shaped fragments. The fragments may be dislocated by muscle spasm.

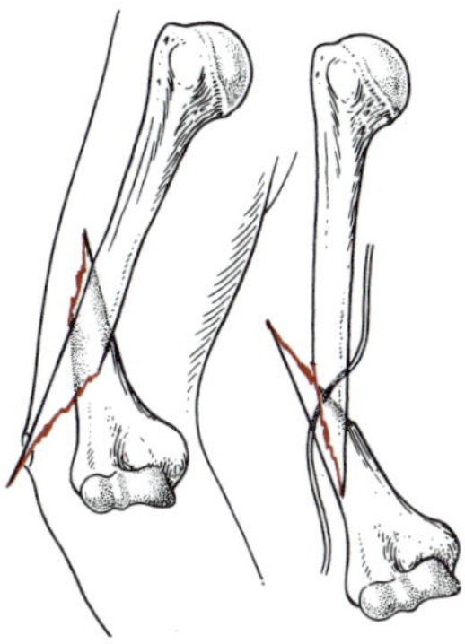

Figure 12–4 Oblique and spiral fractures in the upper arm lead to large dislocations

Source: Schlosser and Kuner, 1980

Clinical Considerations and Diagnosis

The typical symptoms of a fracture—abnormal mobility, malalignment, functional disturbances, and crepitation—are especially pronounced in a fracture of the humeral shaft. The diagnosis is confirmed by radiographic examination.

Medical Management

Unless there is severe dislocation, fractures of the shaft are usually treated nonsurgically. In addition to a hanging cast (see Figure 12–5) or an early functional treatment (following procedures by Specht), a reposition under general anesthesia and immobilization in an airplane splint must be considered. The bony consolidation requires about eight weeks. Surgery is indicated in cases with severe dislocation, insufficient alignment of the fragments in nonsurgically treated fractures, in multiple trauma, or in accompanying lesions of nerves and blood vessels. The techniques used are a plate osteosynthesis or nailing (following procedures suggested by Bunnell), which provides for early exercise stability.

The injured arm is immobilized (2 to 4 days) in a Desault bandage, which extends to the hand, with an extra loop for the hand. After the fifth day, following removal of the bandage, the arm hangs during the day in a large sling, which must not compress the arm. During the night, until the onset of a bony consolidation, the arm is fixed to the trunk by an elastic bandage.

Early and Late Complications

Damage to the radial nerve is a frequent occurrence. The lesion, which might either be immediately apparent or appears only later, may be caused by callus

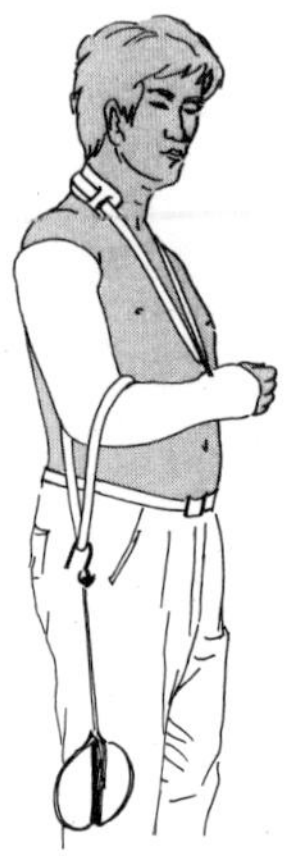

Figure 12–5 Hanging cast as treatment of the upper arm fracture in older adults. Circular movements can be initiated early with this type of fixation.

formation. Neurolysis is mandatory. Pseudarthroses are observed in about 8 percent of cases and are treated surgically.

Symptoms

- usually extensive hematoma formation and edema around the upper arm and elbow
- crepitation of the fracture
- limited range of motion
- pain

Therapeutic Aspects

- edema prophylaxis and support of local blood supply: (1) application of a sling, (2) kneading exercises with a foam ball, (3) active movements, (4) modalities, (5) subsequently, exercises in therapeutic pool, swimming
- mobilization: (1) active movements, (2) active movements with equipment (dumbbells, stall bars, rod, and so on) or with a partner, (3) subsequently, possibly relaxation techniques with applications of ice, (4) subsequently, possibly PNF exercises, use of therapeutic pool, (5) joint mobilization
- strengthening: (1) active movements, (2) active exercises against resistance
- exercising in activities of daily living

Course of Treatment

During the first three weeks, most of the exercises require a special splint. This support helps to prevent the tendency to dislocate, which is typical for fractures of the humerus and usually occurs as a shortening and curvature of the humerus anteriorly and medially.

The use of any rotary mechanism on the still unstable fracture must be avoided, since it will interfere with the healing process and increase the risk of pseudarthrosis.

The therapist may initially be bothered by the crepitation of the fracture, but this is normal and will gradually diminish as the consolidation of the fracture progresses. The patient should knead and press a foam ball during and after the treatment to improve the muscle tone.

Beginning on the first day, the patient moves the hand, which rests in a sling in the Desault bandage. Movements are made in all possible directions. The patient also kneads and presses the foam ball.

The patient is prepared for further therapy through explanations by the therapist and also, if possible, by watching other patients. Beginning on about the third to fifth day, pronation and supination begin.

Sample Exercises

The patient sits upright and the forearm (elbow flexion 90 degrees) is placed in the palm of the therapist's hand. The therapist's other hand (around the humeral condyles) pulls downward along the axis of the upper arm. The patient performs pronation and supination (see Figure 12–6).

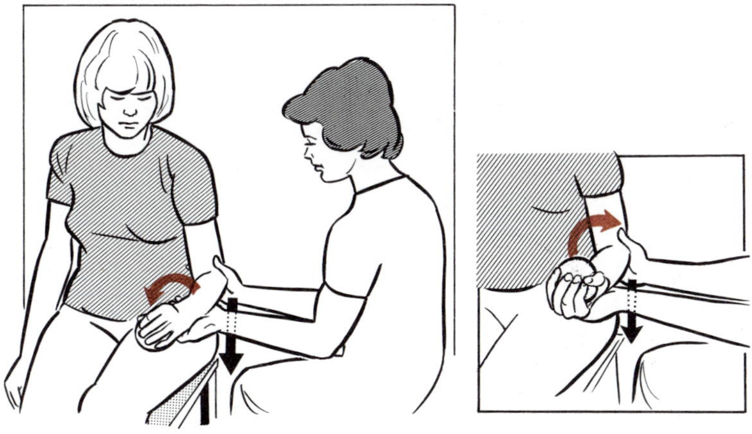

Figure 12–6 a. Pronation and b. supination

Source: Specht, 1976

Also, during this period, elbow flexion and extension are performed. The patient sits upright and the therapist fixes the patient's upper arm against the direction of the dislocation (shortening, medial, and anterior curvature), applying pressure laterally and posteriorly with one hand and applying slight traction along the humeral axis with the other hand. In treating fractures of the left humerus, the therapist's left hand would be the "pressure" hand, the right one the hand applying traction (see Figure 12–7). The patient flexes and extends the arm at the elbow joint, initially from the thigh, later, as the patient approaches the full range of motion, from the side (next to the body).

Finally, in this initial period, swinging exercises should be performed with appropriate equipment: usually 1 kg dumbbells. The patient stands upright swinging the dumbbells in a sagittal plane beside the body. Then, bending slightly forward, with the normal hand on the back of a chair for support, the patient swings the dumbbells transversally in front of the body (see Figure 12–8). All abrupt and circling motions must be avoided.

On about the seventh day, exercises for abduction of the shoulder joint are begun, starting with the elbow in 90 degrees of flexion. The patient sits upright and repeats the second set of exercises for the 3 to 5 day period as described above, except that now the patient's forearm is placed on the therapist's forearm.

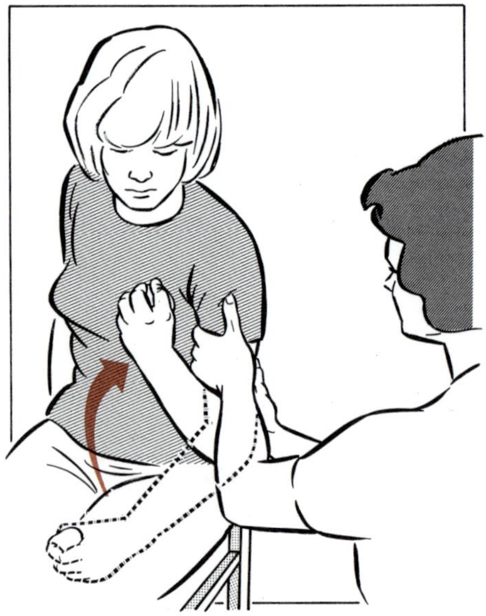

Figure 12–7 Partial extension and flexion

Source: Specht, 1976

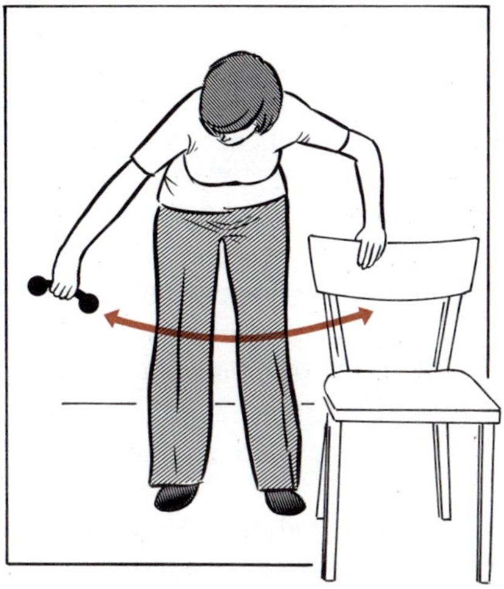

Figure 12–8 Pendulum exercises with a weight

Source: Specht, 1976

Encourage active abduction as far as possible. The patient fracture site must be well stabilized by the therapist (see Figure 12–9).

Beginning on about the tenth to twelfth days, exercises by "dangling" the arm are performed. The patient is standing and trunk is bent forward. Both arms are

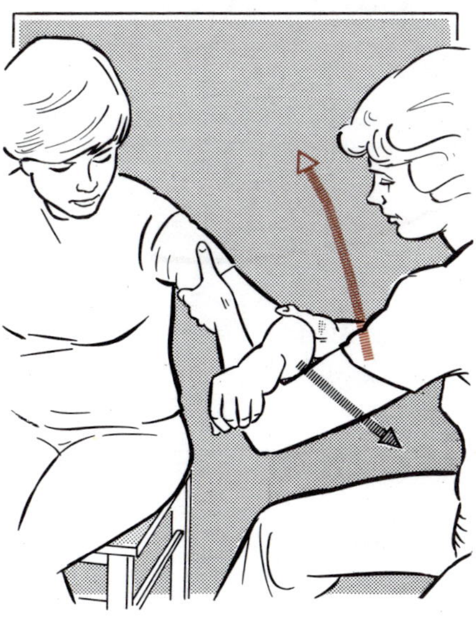

Figure 12–9 Abduction

Source: Specht, 1976

allowed to dangle without weight in a sagittal plane ("wood cutter" exercise) and in a horizontal plane ("floor sweeping") (see Figure 12–10).

After about two or three weeks, when the crepitation has subsided, pulling and pushing exercises are begun. For the pulling position, the patient and therapist sit on stools opposite each other. The therapist grasps both of the patient's wrists dorsally. While both the therapist and the patient lean backward, the patient pulls alternately on the left and the right side. After a few days, the pushing exercises are started, with the therapist and patient now bending forward. In this position, the therapist can feel exactly how much resistance to apply. In this part of the exercise, the therapist grasps both of the patient's wrists ventrally (see Figure 12–11).

Also at this time, exercises at the stall bars can be performed, with the patient alternately pulling and pushing, following the procedures described above (see Figure 12–12).

Finally, in this phase, the patient may exercise using a rod. In either a sitting or standing position, the patient lifts the rod up as high as possible with both arms with elbows extended.

As soon as the fracture is clinically stable, rotating movements are added to the program—particularly movements related to daily routines, such as touching the nose or ear, combing the hair, and so on. Later, supplementing this, exercises in the therapeutic pool and occupational therapy can be started.

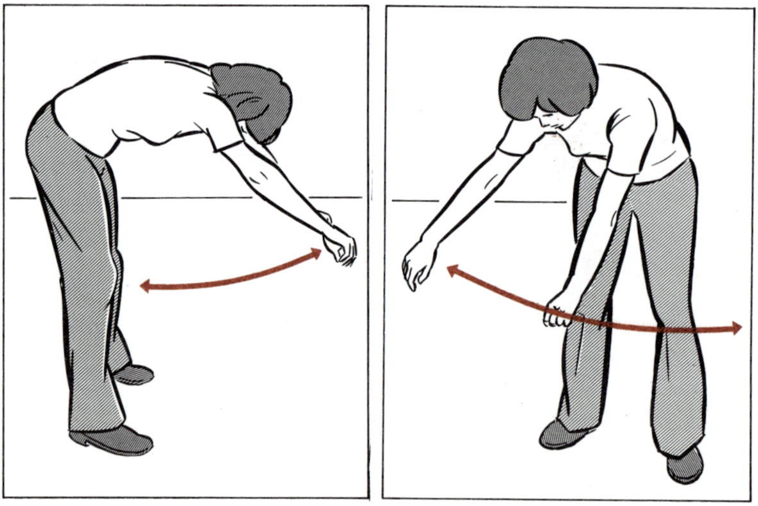

Figure 12–10 a. "Wood chopping" exercise; b. "sweeping"

Source: Specht, 1976

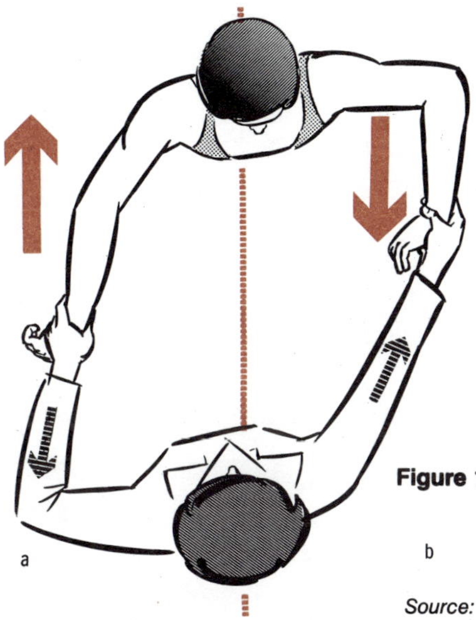

Figure 12–11 Pull and push exercise in sitting position

Source: Specht, 1976

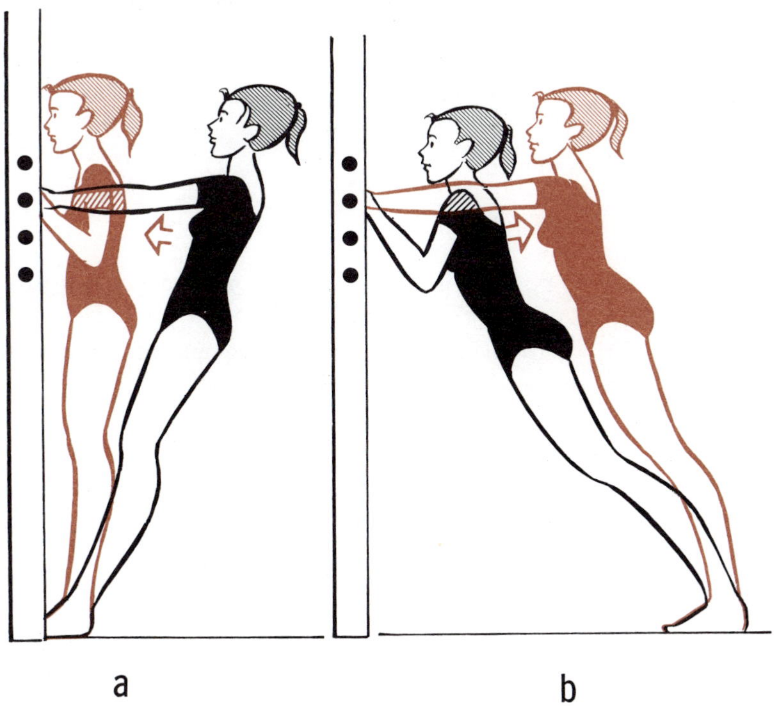

Figure 12–12 a. Pull and b. push exercise at the wall bars

Treatment Using a Hanging Cast

During the first four weeks, the injured arm is immobilized in a hanging cast. The mobility of the fingers is usually normal. Range of motion of the shoulder joint is limited by the cast.

Symptoms

After removal of the cast, the following symptoms are observable:

- tendency of the arm to swell and be discolored (bluish red), especially distally
- possibly residual hematoma
- pitting edema
- limited range of motion of hand and elbow
- atrophy and weakness of all the muscles of the arm
- limited ability to perform daily activities
- pain

Therapeutic Aspects

During the immobilization phase, the following therapeutic procedures are applicable:

- edema prophylaxis
- mobilization: active movements
- strengthening: active movements

After removal of the cast, the following procedures are applicable:

- edema prophylaxis: (1) positioning, (2) compression bandages, (3) isometric and dynamic exercises, (4) therapeutic pool, (5) ice applications
- mobilization: (1) active movements with and without equipment, (2) relaxation techniques with ice applications, (3) therapeutic pool, (4) joint mobilization
- strengthening: (1) dynamic exercises against resistance (PNF), (2) therapeutic pool
- training in activities of daily living

Course of Treatment

In the first four weeks, the patient is instructed to exercise fingers and shoulder, following procedures by Poelchen. The shoulder exercises are especially important to maintain or restore the full range of motion.

In about the fifth week, the cast is removed. In case of edema, the arm must be elevated temporarily and wrapped with an elastic bandage. Mobilization of the hand and elbow joint is best done by exercises in the therapeutic pool. For mobilization, relaxation techniques should be applied in combination with ice applications, active movements of the arm, and the use of equipment (ball, rod, and so on).

As soon as the fracture is stable, resistive exercises and training in activities of daily living are indicated. These should be followed by swimming and occupational therapy to conclude the rehabilitation program.

Sample Exercises

In the early phase of mobilization and strengthening, with the hanging cast in place, the patient assumes a standing position, bending forward. With the therapist supporting the patient's arm during the first days, the patient performs the following arm exercises:

- swinging backward and forward (sagittal plane)
- swinging to the sides (frontal plane)
- as far as possible, making circling motions while maintaining moderate rotation

Next, with the patient in the same position as above and with the arm lightly supported by the therapist or by the patient's normal hand, the patient

- moves the arm into shoulder flexion, then to the sides in shoulder abduction
- moves the arm passively in a circular pattern in both directions

All of the above exercises should be started with the patient pressing a foam ball or holding the fingers in a tight fist to build up the tone distally.

Surgically Treated Fracture of the Humeral Shaft

The arm is either wrapped in an elastic bandage or placed in a cast and elevated on a wedge pillow (see Figure 12–13).

Symptoms

After removal of the bandage or cast, the following symptoms are observable:

- postoperative hematoma and edema
- possibly elevated local temperature
- scar adhesions to the adjacent tissue

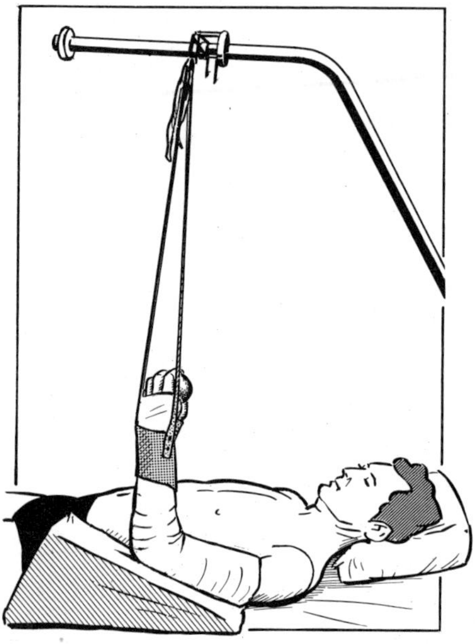

Figure 12–13 Arm positioning possibility

- pain
- possibly motor or sensory deficits, especially in areas innervated by the radial nerve

Therapeutic Aspects

- edema prophylaxis: (1) elevation, (2) compression bandage, (3) changing positions, (4) isometric and dynamic exercise, (5) ice
- mobilization: (1) active movements, (2) possibly relaxation techniques with ice applications, (3) use of therapeutic pool, (4) joint mobilization, (5) friction massage of scars
- strengthening: (1) active movement against resistance (PNF)
- training in activities of daily living

Course of Treatment

During the first postoperative days, to avoid edema, it is advisable to wrap the arm in an elastic bandage and elevate it temporarily. As early as the second day, the physical therapy can begin with dynamic exercises for the full arm. Most likely the therapist will have to support the arm initially. These first exercises should be done in a supine or sitting position with light equipment (cane, tennis ball).

On the tenth day, if there are no complications in the healing process, the stitches are removed. Afterward, the patient is allowed to bathe the arm and to exercise in the therapeutic pool. If persistent contractures have developed in spite of early mobilization, they can be treated with relaxation, ice applications, and joint mobilization. As soon as the fracture is clinically stable, exercises against resistance and occupational therapy are added to the treatment. Due to the early achievement of exercise stability, the patient regains the complete range of motion fairly quickly.

Sample Exercises

The following exercises are performed in the early phase of mobilization and strengthening. In the first set of exercises, the patient is in a supine or sitting position, with the upper arm resting on a flat foam pillow (from shoulder to elbow) or on a small treatment table and with the forearm elevated at a right angle:

- making all active movements of the fingers and wrist
- elbow flexion with a tight fist and elbow extension with extended fingers and wrist
- the above movements, if possible, with supination and pronation, combining elbow flexion with supination, elbow extension with pronation

The above exercises may also be performed using lightweight equipment, such as a foam ball, finger dumbbells made of wood, and so on, to achieve more intense distal tone.

The following exercises are done while in a side-lying position on a mat with the injured arm elevated, using a ball:

- with the hand placed on the ball, rolling it with the best possible flexion of the elbow from in front of the body straight forward, then up and down, toward various points on an imaginery semicircle, extending the elbow as much as possible, then moving the ball back to the body with flexed elbow
- in the same position, moving the ball in front of the body in wide circles (both directions), trying to achieve the best possible flexion of the elbow close to the body and the best possible extension away from the body

Complications

If the radial nerve is paralyzed, the following therapeutic procedures are included in the exercise program:

- support and prevention of contractures by a radial splint (correction of a wrist drop) (see Figure 12–14)
- support of the blood flow
- electrotherapy—provided that no metal implants, such as screws or plates, interfere with the electrical field
- short, quick ice applications in combination with exercise techniques to support the innervation
- training in activities of daily living

DISTAL HUMERUS FRACTURES

Mechanism of Injury

In distal fractures of the humerus among adults, direct trauma, especially falling on the elbow, is most common. In children, indirect trauma, such as falls on the hand, with the elbow extended, is most common.

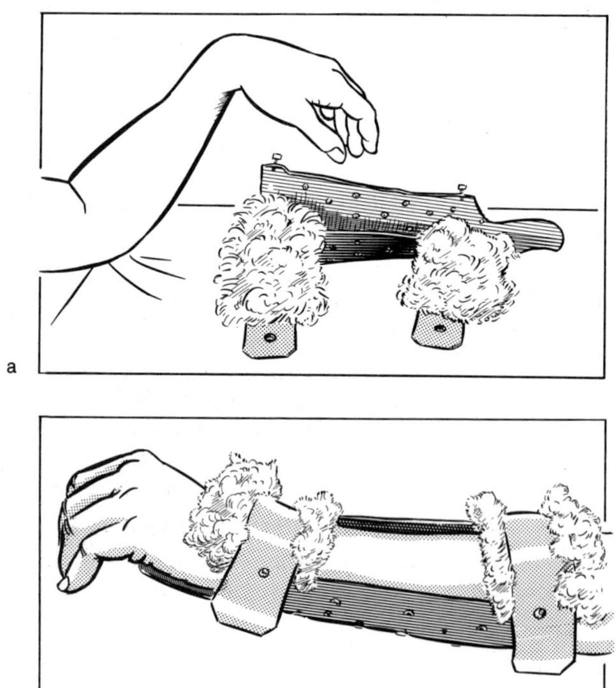

Figure 12–14 Splint for radialis paresis

Types

Fractures of the distal humerus occur in various shapes: transverse fractures of the distal shaft; X- and Y-shaped fractures extending into the joint; isolated separation fractures of the condyles; and, especially in children, transverse supracondylar fractures (related to flexion or extension of the elbow) and fractures of the head of the radius (see Figure 12–15).

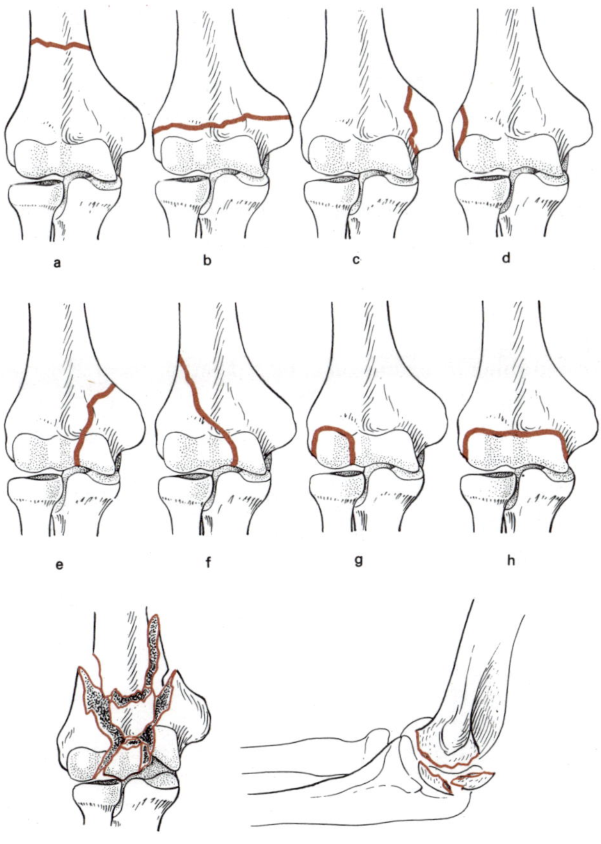

Figure 12–15 Types of distal humerus fractures: a. transverse fracture of the distal humerus; b. transverse fracture through the condyles; c. epicondylar fragmentation on the ulnar side; d. epicondylar fragmentation on the radial side; e. fracture of the condyle on the ulnar side; f. fracture of the condyle on the radial side; g. shearing off of the condyle on the radial side; h. fracture of the trochlea; i. multiple fractures of the upper arm; k. side view of a fracture caused by a shearing force.

Source: Schlosser and Kuner, 1980

Clinical Considerations and Diagnosis

The diagnosis is based on observations of pain, swelling, bony malalignment, abnormal mobility, and the results of radiographic examination. Simultaneous injuries of nerves and vessels must be considered.

Medical Management

Slightly dislocated pericondylar and nondisplaced intra-articular fractures in adult patients are treated nonsurgically with a bivalved cast for four weeks, with the elbow flexed to 90 degrees. Afterward, physical therapy is started, at first, with the cast in place. More severe intra-articular fractures in older patients are also treated nonsurgically.

The fracture is reduced by an olecranon wire extension (see Figure 12–16). Afterward, it is immobilized in a cast. Surgical treatment is indicated in compound fractures and in fractures that cause disruptions of the articular congruency. Surgical treatment is also indicated in fractures that involve vessels or nerves. Screws and ¼ inch tube plates are used to stabilize the osteosynthesis and permit early active physical therapy.

Fractures in infants are treated either by extension or by reposition under general anesthesia and immobilization. The alignment has to be correct to prevent any malalignment during the growth period. This is especially crucial in a lesion of the epiphyseal growth plate (see Figure 12–17). A reduced but unstable fracture has to be secured by crossed drilling wires. The blood supply must be observed carefully.

Displaced fractures that involve the growth plate have to be treated surgically. Exact alignment and wire fixation are necessary. The blood circulation must not be

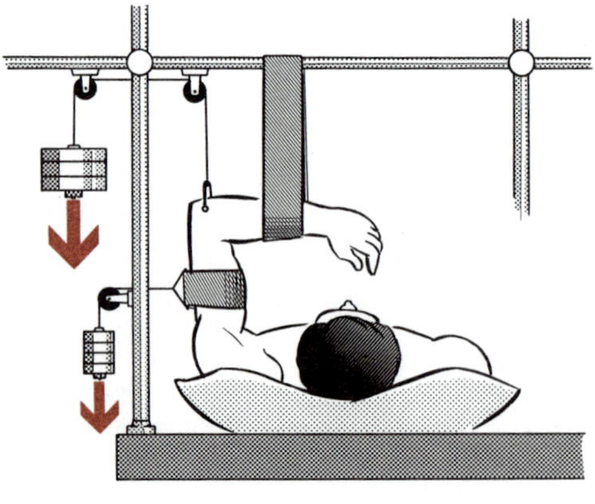

Figure 12–16 Extension by Baumann through the olecranon in order to get a vertical force

Source: Schlosser and Kuner, 1980

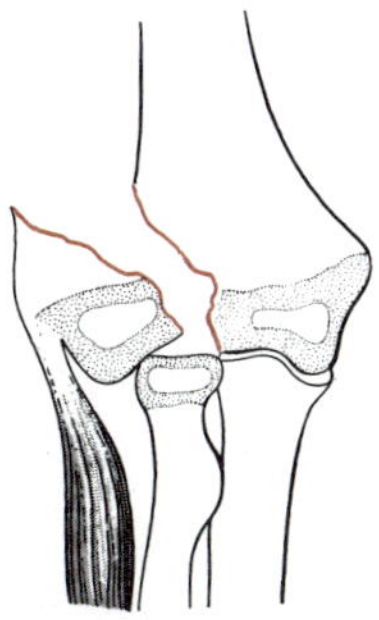

Figure 12–17 Fragmentation of the distal humerus on the radial side

Source: Schlosser and Kuner, 1980

constricted by bandages; any such interference with the blood supply increases the risk of ischemia (see subsequent section on Volkmann's contracture). Also, the danger of a myositis ossificans around the joint must be guarded against.

Early and Late Complications

Lesions of the ulnar and medial nerve can occur. Also, a limited range of motion may result from joint degeneration. A late complication may be paralysis of the radial nerve as a result of callus formation.

Nonsurgically Treated Supracondylar Fracture of the Humerus

Symptoms

After removal of the cast, the following symptoms are usually observable:

- remaining hematoma and a general tendency toward discoloration of the arm when it is hanging down
- edematous tissue
- tendency of the arm to swell, especially in areas distal to the fracture
- limited range of motion of the joints of hand and elbow
- atrophy and weakness of the muscles of the arm
- disability in performing daily activities
- pain

Therapeutic Aspects

- edema prophylaxis: (1) positioning, changing positions, (2) compression bandage, (3) isometric and dynamic exercises, (4) ice, (5) hydrotherapy

- mobilization: (1) active movements, (2) relaxation techniques with ice applications, (3) use of therapeutic pool
- strengthening: (1) exercises against resistance (PNF), (2) use of therapeutic pool
- training in activities of daily living

Course of Treatment

To prevent a stiff shoulder during the immobilization period in the first three weeks, the patient exercises following previously described guidelines for prophylactic therapy for patients with injured arms. During the fourth week, the cast is split and vigorous physical therapy is started. If there is edematous swelling (see Figure 12–18), the arm must be elevated, treated with ice, or wrapped in elastic bandages. Also, position changes must be exercised. At this point, intensive mobilization of the hand and elbow can be started. The increase in range of motion is facilitated by applications of ice.

As soon as the fracture is stable enough, exercises against resistance and training in daily activities can be started. Occupational therapy concludes the therapeutic program.

Sample Exercises

The following exercises are used in the late phase of mobilization and strengthening to increase the range of motion. In the first set, the patient, lying in a supine

Figure 12–18 Positioning bench to reduce swelling

position, moves the forearm from extension-pronation into flexion-supination, using relaxation techniques at the end of the range of motion and securing the achieved flexion-supination by repeated contraction.

In the second set of exercises, PNF techniques are used in both diagonals, working each from elbow extension to flexion and applying relaxation techniques at the end of the elbow flexion: (1) slowly returning, holding, relaxing, or (2) contracting, then relaxing, and improving the result by repeated contractions, slow returns, and a final slow return with holding.

In the third set of exercises, the patient lies in a supine position, holding a ball or rod with both hands, then flexing as close as possible toward the head, trying to extend the range at the end of the flexion with a relaxation technique, then strengthening with repeated contractions.

In the final set of exercises, the patient is in a supine position, with the upper arm resting on a pillow or wedge and the forearm elevated at a right angle. Holding a pair of dumbbells (weight depending on the patient's ability and limitations), the patient moves them through an arc of motion toward the shoulder. The patient relaxes at the end of the movement allowing the elbow to be passively flexed by the weight of the dumbbells. This same exercise may also be done in a sitting position with the arm placed on a small treatment table.

Surgically Treated Supracondylar Fracture of the Humerus

The patient's arm (in a bivalved cast with elbow flexion) is placed on a pillow in abduction.

Symptoms

After removal of the cast, the following symptoms are observable:

- residual hematoma and postoperative edema
- scar adhesions to the adjacent tissue
- edematous tissue, especially around the scar
- limited range of motion of hand and elbow
- pain

Therapeutic Aspects

The therapeutic procedures for a surgically treated supracondylar fracture of the humerus are the same as those for nonsurgical treatment, except that scar treatment to increase the blood supply might be added.

Course of Treatment

For the first ten days, the injured arm rests on an abduction pillow, usually in a bivalved cast. The patient exercises intensively following previously described procedures for prophylactic treatment of patients with injured arms.

On about the eleventh day (possibly earlier if there is sufficient stability and the soft tissue is in good shape), physical therapy is initiated. If necessary, edema and a disturbed blood supply should be treated by elevation, ice applications, compression, and by changing positions. After the healing of the soft tissue, further supportive measures are added to the treatment, such as active exercises in a whirlpool.

After the eleventh day, at the latest, range of motion exercises are begun in conjunction with ice applications. Also, the muscles are strengthened; this is done by dynamic exercises with lightweight equipment and without any shearing load, since there is still little stability. The joints of the elbow, hand, and fingers can be moved separately or together using various complex motions.

In the following weeks, as soon as the fracture shows bony consolidation, exercises against resistance and occupational therapy are added to the program.

Sample Exercises

The following exercises are used in the early phase of mobilization and strengthening of the hand:

- active movements of the wrist in all directions: flexion, extension, radial deviation, ulnar deviation, circling in both directions
- moving from a position of closed fist, flexed wrist, supinated forearm and elbow flexion to a position of extended fingers and wrist, pronated forearm and elbow extension
- with hands on a round pillow, rolling the pillow away from the body in elbow extension and wrist flexion, then back toward the body in elbow flexion and wrist extension
- with the hand on a ball, rolling the ball backward and forward, to both sides, in diagonal directions, and in a circle
- doing the above exercise bilaterally with a gymnastic ball

The following set of extension exercises is performed for mobilization and strengthening of the elbow. First, in the early phase:

- in a supine position or sitting with the upper arm placed as in the first exercise described earlier for surgically treated fractures of the humerus, combining strictly dynamic elbow extension exercises with finger-hand extension, plus pronation, or exercising in elbow extension, plus pronation, while pressing a foam or tennis ball

- in a supine position, with parallel arms and both hands holding a cane (in pronation and supination), flexing with the cane toward the face and extending it back toward the legs
- doing the same exercise with extension toward the ceiling
- in the same position, moving the cane in a wide circle across the body (both directions)

In the late phase:

- as in the first exercise described earlier for a nonsurgically treated fracture of the humerus, extending the elbow extension-pronation by relaxation techniques and then strengthening by repeated contractions
- using PNF exercises in both diagonals, each extending from a flexed to an extended elbow, repeating the second exercise described earlier for a nonsurgically treated supracondylar fracture of the humerus
- using PNF exercises, working bilaterally and symmetrically with extended elbow in a shoulder flexion-abduction-external rotation, then resisting the normal arm and moving the injured arm through ''stretch'' and ''approximation'' positions to the final position of the diagonal, while maintaining the maximal elbow extension or stimulating it by approximation
- while in a standing position, moving dumbbells from 90 degrees elbow flexion to extension-supination and extension-pronation, letting the weight of the dumbbells themselves accentuate the end phase of the movement
- with dumbbells, isometric contractions of all arm muscles, relaxing, and actively stretching

Chapter 13

Injuries of the Elbow Joint and Forearm

LIGAMENTOUS INJURY AND SUBLUXATION

The elbow joint is a combination of three compartmented joints: the humeroulnar joint, the humeroradial joint, and the radioulnar joint. The elbow joint acts as a hinge between the upper arm and forearm, allowing for rotation of the forearm around the longitudinal axis.

Mechanism of Injury

Elbow injuries are as variable as the mechanism of the joint itself. The principal cause is a fall on the arm or hand or a rotational force to the forearm.

Types

In addition to contusions, comminutions, and sprains, capsular tears and ligamentous ruptures can occur. If the impact is very severe and the ligaments rupture, the injury can result in a subluxation of the joint between the radius and humerus or in a subluxation of the ulnar head. The combination of a subluxation of the radial head and an ulnar fracture is called a Monteggia fracture.

Clinical Considerations and Diagnosis

Pain, swelling, and a limited range of motion are characteristic of less severe elbow injuries. Ruptured ligaments result in instability and subluxation; a subluxation results in deformation and loss of motion.

Medical Management

Minor injuries are treated functionally and nonsurgically. Partial and complete ligament ruptures require immobilization, or even surgery to repair the lateral collateral or the annular ligament. The ulnar subluxation is usually treated nonsurgically, with reduction and immobilization for four weeks.

Simultaneous fractures of the olecranon and the coronoid process are treated best by internal screw fixation. Unstable ligaments require sutures. Old subluxations have to be reduced surgically and treated by surgical repair of the ligaments. Surgery is also indicated in simultaneous lesions of vessels and nerves and in subluxation fractures.

Early and Late Complications

Lesions of vessels and nerves can occur, and subluxations may remain untreated.

Physical Therapy for Surgically Treated Subluxation of the Elbow

The patient is treated with a bivalved cast, which is slightly elevated and abducted on a pillow.

Symptoms

After removal of the cast, the observable symptoms are the same as in a surgically treated supracondylar fracture of the humerus, except that the tendency to swell around the elbow and the distal area is even more pronounced. Atrophy and weakness of the muscles are also apparent.

Therapeutic Aspects

The conservative therapeutic procedures are basically the same as for a surgically treated supracondylar fracture of the humerus with the addition of treatment for scar tissue.

Course of Treatment

During the three weeks of immobilization, the patient should work independently, following previously described guidelines for prophylactic therapy for patients with injured arms. Isometric exercises are allowed to the extent possible with the cast.

After removal of the cast, edema prophylaxis and exercises to support the local blood flow and to strengthen and mobilize the muscles are started, and training in activities of daily living begins. The therapy should be applied very carefully, since the elbow joint is very delicate. Ice application helps to increase the range of motion and prevent inflammation.

Sample Exercises

The following exercises are used in the late phase of mobilization and strengthening. First, with the patient in a supine position, a cane is placed diagonally (lower right side to upper left side) into the two hands. The injured hand (upper hand) grasps the cane in supination, the other hand in pronation. The patient pushes the cane against the therapist's resistance to the left and up until the elbow is extended and supinated as much as possible. At the end of the movement, the patient does isometric contractions using relaxation techniques or "rhythmic stabilization," then pushes the cane further in the direction of the final movement. The entire exercise is then repeated slowly in reverse, with a final holding. Finally, the same set of exercises is repeated, using the injured hand as the lower hand.

In the second set of exercises, while the patient stands at a rail, the head and body are pulled as a single unit with flexion-supination of the elbow and forearm to the rail. The patient then grasps the rail in pronation and pushes the body away from the rail until the arms are in elbow extension and forearm pronation. The body remains upright throughout the exercise.

In the final phase of exercises, the patient trains in activities of daily living—combing, washing hair, putting on shirt, etc.

OLECRANON FRACTURES

Mechanism of Injury

The fracture of the olecranon is caused by a fall on the flexed elbow or by a direct impact. The strong pull of the triceps muscle may cause a dislocation of the proximal fragment.

Types

Fractures of the olecranon are classified as transverse, oblique, and comminuted (see Figure 13–1). Frequently, the coronoid process of the ulna is affected.

Clinical Considerations and Diagnosis

The arm cannot be extended against resistance; abnormal contour and palpable depression are present if there is a dislocation.

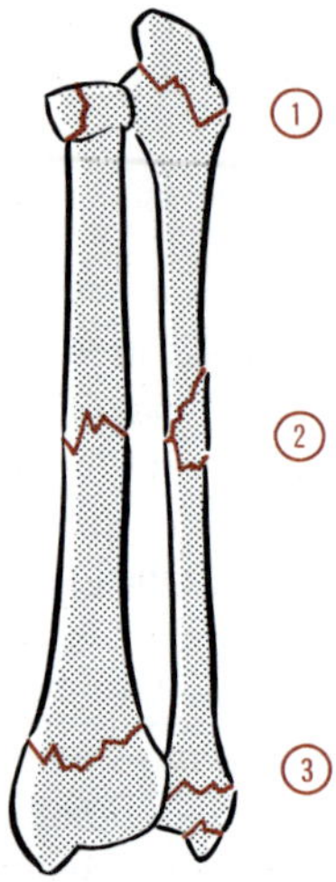

Figure 13–1 Fractures of the forearm: 1. olecranon tear and radial head-chisel fracture; 2. radius fracture, transverse ulna fracture and flexion fracture; 3. distal radius and ulnar head fracture.

Medical Management

Dislocated fractures of the olecranon require surgical treatment. In such treatment, the use of a tension-bent wire has proved to be particularly beneficial. In the following nonsurgical treatment, a cast is applied for about four weeks. This is combined with isometric and active exercises of the shoulder joint.

Complications

If the ulnar nerve is damaged, the fracture has to be reduced and the nerve must be freed and its integrity checked. Following nonsurgical treatment, pseudarthrosis or diastasis may occur. There may be insufficient extension of the elbow.

Physical Therapy for Fracture of the Olecranon Treated With Tension-Bent Wiring

The patient's arm is either wrapped in an elastic bandage or placed in a bivalved cast and elevated on a wedge pillow.

Symptoms

- residual hematoma, postoperative edema
- edematous tissue
- scar adhesions
- limited range of motion of the elbow, sometimes also the wrist and forearm
- pain

Therapeutic Aspects

The conservative therapeutic procedures are the same as for surgically treated supracondylar fractures of the humerus with the addition of treatment for scar tissue.

Course of Treatment

During the first three postoperative days, the patient actively exercises fingers, wrist, and shoulder joint. The biceps muscle is worked with isometric contractions. Beginning on the fourth day, edema prophylaxis is combined with active exercises in elbow flexion and extension and in supination and pronation.

Beginning on the eleventh day, after the wound has healed completely, ice application can be used to improve the mobilization of the elbow. Exercises in the whirlpool can then be added.

After about the third week, the patient participates in occupational therapy. The physical therapy is expanded to include more complex exercises and training in activities of daily living. Following bony consolidation of the fracture, exercises against resistance with equipment are allowed.

Sample Exercises

The following exercises are used in the early phase of the mobilization. For the first three days the bivalved cast is opened and the arm is held by the therapist in 90 degrees of flexion (one hand at the wrist, the other one above the olecranon):

- moving the fingers and wrist actively, pressing a foam ball or kneading theraputty
- with the elbow in the same position, making all movements of the shoulder while the therapist supports the arm
- isometric contractions of the elbow extensors

Beginning on the fourth day, in the same position as above:

- elbow flexion to point of tension
- elbow flexion with supination and pronation
- raising the arm with a tight fist toward the top of head, each time bringing it down again beside the body

This exercise may subsequently be expanded, using PNF diagonals, including those in conjunction with the elbow.

Beginning on the eleventh day, elbow flexion and extension are practiced with a ball:

- bilaterally, with the ball in both hands, moving ball toward the body in flexion, away from the body in extension
- with the injured arm in circumduction just below the surface of the water, moving it to the side (in abduction) and back into circumduction, so that the fingers touch the opposite shoulder
- with the injured arm held downward next to the body and the elbow in the best possible extension, flexing the elbow with a closed fist, supination, and back into extension-pronation
- exercising with PNF, both diagonals, including the elbow

RADIAL HEAD FRACTURES

Mechanism of Injury

Fractures of the radial head are caused mainly by direct force.

Types

Chisel fractures, as well as transverse and comminuted fractures, can occur.

Symptoms, Clinical Considerations, and Diagnosis

Symptoms include swelling, pain, and a limited range of motion, especially in pronation and supination. Sometimes a tomogram can be useful in diagnosis.

Medical Management

A fracture of the radial head, with or without a slight dislocation, requires immobilization for 2 to 3 weeks, followed by exercise therapy, especially active exercises. A severe dislocation has to be reduced surgically and fixed by screws. Small fragments might have to be removed. A crushed radial head must be resected; this should be avoided, however, if possible, in adolescents.

Complications

Injuries of the deep branch of the radial nerve, paralysis of the extensors, and limited pronation-supination are possible complications.

Physical Therapy for Nonsurgically Treated Chisel Fracture

Symptoms

The injured arm is immobilized in a splint or, if there is no dislocation, is wrapped in an elastic bandage. The arm is elevated and abducted on a pillow.

The following symptoms are observable:

- possibly residual hematoma
- edematous tissue
- limited range of motion, especially of pronation-supination
- pain and tendency to keep the injured arm still

Therapeutic Aspects

The therapeutic procedures are basically the same as in a nonsurgically treated fracture of the humerus.

Course of Treatment

Beginning on the first day, the patient exercises following previously described guidelines for prophylactic treatment of patients with injured arms. On the third day at the earliest, physical therapy is started. The swelling is treated with elevation and ice applications and by changing positions. The resorption of large hematomas can be accelerated by modalities.

The mobilization of all joints is initiated with active exercises, possibly in combination with ice applications. At first, the exercises are selective, then they are combined in more complex movements. Exercises in the whirlpool or therapeutic pool are started at this time.

Following complete consolidation of the fracture, exercises against resistance, training in daily activities, occupational therapy, conclude the therapeutic program.

Sample Exercises

The following exercises are used in middle to late phases of mobilization and strengthening and in training in daily activities. In the first set of exercises, the patient is in a supine position, with the elbow in 90 degrees of flexion:

- selective exercising of supination-pronation, creating distal tension by making a tight fist or pressing a foam ball
- using a screwdriver to insert a screw into a piece of wood then unscrewing it.

In the second set of exercises, the patient sits at a treatment table with the forearm placed on the table (80 to 90 degree elbow flexion) and the hand holding a rod across the edge of the table, then turns the rod back and forth causing the forearm to supinate and pronate.

- exercising supination and pronation
- doing the same exercise with dumbbells
- putting nuts onto a bolt, etc.

Finally, the following exercises may be performed while training in daily activities:

- turning a doorknob
- turning a key
- pouring water out of a pot into a cup and pouring it back
- wrapping yarn or a string

Physical Therapy for Surgical Resection of the Radial Head

Symptoms

The injured arm is placed in a cast (elbow flexion 90 degrees) and elevated on a wedge pillow. The following symptoms are observable:

- residual hematoma
- postoperative swelling and tendency to swell distally
- edematous tissue

- scar adhesions
- limited range of motion
- pain

Therapeutic Aspects

The conservative therapeutic procedures are the same as for a surgically treated supracondylar fracture of the humerus with the addition of treatment for scar tissue.

Course of Treatment

Starting on the first day, the patient is instructed in exercises, following previously described guidelines for the prophylactic treatment of patients with injured arms. After the wound has healed completely, the treatment is the same as that beginning with the third day of the nonsurgical treatment program.

Sample Exercises

Depending on the phase, the following previously described exercises may be performed:

- exercises for a surgically treated fracture of the humeral shaft
- exercises for a supracondylar fracture of the humerus
- exercises for a subluxation of the elbow
- exercises for a nonsurgically treated fracture of the radial head

Complications in Injuries of the Elbow Joint

Injuries of the elbow, especially subluxations of the elbow, are frequently associated with extensive soft-tissue damage. They carry the risk of myositis ossificans and, especially in combination with fractures of the distal humerus, an ischemic muscle contracture caused by compressed blood vessels. Because of its position close to the joint, the ulnar nerve is at risk of malalignment after fractures close to the elbow, hypertrophic callus formation, or because of a subluxation of the elbow.

ISCHEMIC CONTRACTURE

Mechanism of Injury

An ischemic contracture (Volkmann's contracture) is caused by a permanent pressure on the inside of the elbow. This pressure can result from hemorrhages into the tissue, swelling and compression of blood vessels, lesion of the brachial artery by bony fragments, or compression by a tight cast. The generally increased pressure decreases the blood supply, creating tissue necrosis. If the destruction of tissue exceeds the tissue's regenerative ability, it will result in severe complications.

Clinical Considerations

The acute disturbance is characterized by edematous swelling, decreased circulation, alterations of the vascular permeability, and loss of mobility.

Symptoms

The symptoms include pain and discoloration of the forearm, including the hand. The pulse is not palpable. Six hours without oxygen are enough to cause destruction of the muscle. In the progressive stage, the wrist is fixed in flexion; extension of the fingers is impossible. The flexor muscles are scarred and contractures result. The ulnar and median nerve can show deficits due to scar compression.

Medical Management

The treatment consists of immediate elimination of the pressure causing the contracture. Early intervention is necessary for successful treatment. Elevation of the arm supplements the treatment. Application of heat is contraindicated.

A pronounced contracture requires intensive therapy, including positioning in traction and physical and occupational therapy. If the result of nonsurgical treatment is inadequate, extensive surgery must be considered to remove the scar tissue and to elongate the flexor tendons.

MYOSITIS OSSIFICANS

Myositis ossificans is frequently observed after fractures and subluxations of the elbow and must therefore be considered in the therapeutic program. Infants and adolescents, whose bone cells are still very active, are primarily affected by this complication.

Mechanism of Injury

Myositis ossificans may be caused by hematomas and injuries to blood vessels in muscles and other tissues. However, incorrect treatment may also be a cause—

especially massages of the joint area or nonprofessional manipulations, such as in repeated attempts to reduce a fracture or subluxation or passive range of motion.

Types

There are many gradations of myositis ossificans ranging up to a completely immobilized joint.

Symptoms, Clinical Considerations, and Diagnosis

The symptoms are pain, inhibited motion, locally elevated temperature, swelling, and deformation. An attempt to tense the involved muscle may cause spasms. Radiographic examination can reveal periarticular calcifications.

Medical Management

The treatment starts with prevention. Frequent forceful manipulations have to be avoided. Drugs such as diphosphates might be used. In the case of radiographically mature bony formation, an arthrolysis must be done. Reoccurrence is frequent.

INJURIES OF THE ULNAR NERVE

Injuries of the ulnar nerve are the most frequent of all nerve injuries because of the ulnar nerve's position close to the surface at the medial humeral epicondyle.

Mechanism of Injury

The ulnar nerve is affected primarily in fractures of the medial humeral condyle. Acute compressions are caused by swelling, hematoma, or direct pressure from the fragment. In elbow injuries caused by hypertrophic callus formation, malalignment, or scar tissue, paralysis of the ulnar nerve will become obvious within the first year. The onset of the paralysis is insidious, with mild sensory disturbances and atrophy of the intrinsic hand muscles.

Clinical Consideration and Diagnosis

The paralysis of the ulnar nerve is revealed by characteristic sensory and motor deficits. The condition can be functionally tested by holding a piece of paper between the thumb and second metacarpal.

Medical Management

The treatment of an acute paralysis is immediate decompression, reduction of the fragments (if that has not already been done), and perhaps a deviation of the ulnar nerve to the volar side of the arm.

Unless there is irreparable damage by compression or disrupted continuity (lacerations by bone fragments), recovery of the nerve can be expected within several weeks or months. A splint is necessary in persistent deficits to support and prevent contractures. The typical chronic clinical result is the clawhand. Involvement of the brachial plexus must be ruled out as a differential diagnosis.

Physical Therapy for Paralysis of the Ulnar Nerve

In the case of a paralyzed ulnar nerve, the treatment program must be extended to include neurologic aspects and procedures. Usually, the following procedures are recommended:

- support and contracture prevention with a splint (see Figure 13–2)
- improvement of local blood flow by kneading of theraputty
- electrotherapy in cases where there is no implant material (plates, screws) that might interfere with the electrical field
- short, quick ice applications in combination with exercises for the paralyzed muscles; as soon as possible, PNF techniques
- training in activities of daily living

FRACTURES OF THE FOREARM

Radial and Ulnar Shaft Fractures

Mechanism of Injury

Fractures of the shafts are usually caused by a direct impact, for example, by a blow or fall on a sharp edge.

Types

Fractures of the bones of the forearm occur in various shapes. Either the radius or ulna may be fractured or both bones might be involved. A Galeazzi fracture is a radial shaft fracture combined with a subluxation of the ulna in the wrist joint.

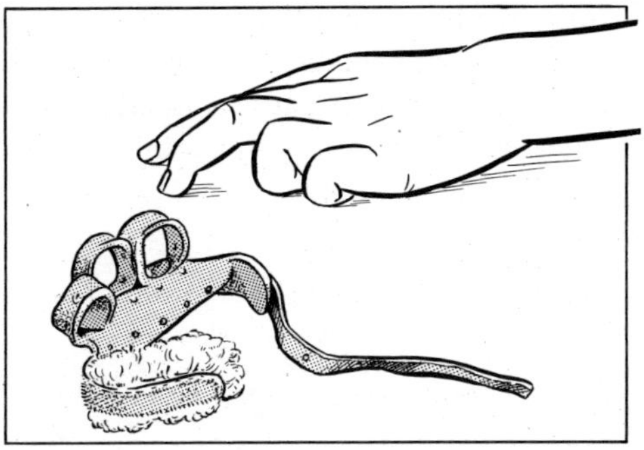

a

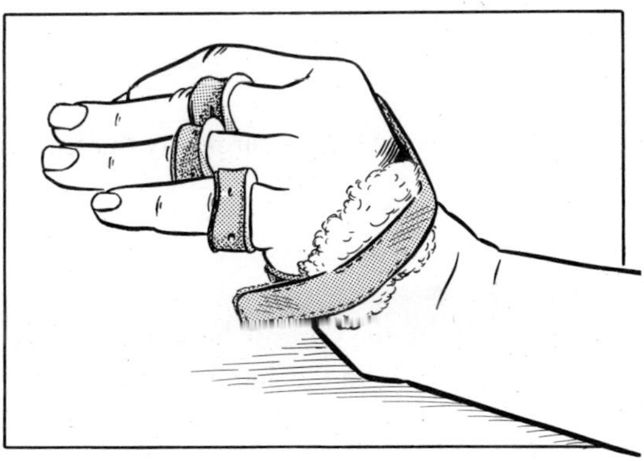

b

Figure 13–2 Positioning splint for ulnaris paresis

Symptoms, Clinical Considerations, and Diagnosis

The main symptoms are pain, especially in turning from supination and pronation, and abnormal mobility. The diagnosis of only one fractured bone without dislocation and crepitation is possible only by radiographic examination.

Medical Management

In the case of incomplete or nondisplaced fractures and in infant fractures (possibly after a reduction), the treatment is nonsurgical. These fractures have to be immobilized in a cast for several weeks. The upper arm and the hand up to the finger joints must be included in the cast for anatomical and mechanical reasons.

In infants and adolescents, fractures of the forearm with intact periosteum (greenstick fractures) are relatively easy to reduce. They require immobilization in a cast for about four weeks.

The remaining fracture types require a plate osteosynthesis. This is especially true for fractures of only one bone. Otherwise, the intact bone may interfere with the consolidation of the fractured bone and ultimately cause a pseudarthrosis. The osteosynthesis also prevents complications from a long immobilization period, which are mainly limitations in pronation-supination.

Complications

Pseudarthroses and limited supination-pronation are possible complications.

Physical Therapy for Surgically Treated Monteggia Fracture

Symptoms

The symptoms are the same as those observed after a surgically treated supracondylar fracture of the humerus.

Therapeutic Aspects

The conservative therapeutic procedures are the same as those for a surgically treated supracondylar fracture of the humerus with the addition of treatment for scar tissue.

Course of Treatment

During the first ten days, the patient exercises under supervision and independently, following previously described guidelines for prophylactic treatment of patients with injured arms. On about the eleventh day, the cast is removed to permit special physical therapy for the hand and elbow joints. Except for pronation and supination, dynamic muscle work can be started, with the patient either in a supine position or sitting at a treatment table. Initially, it might be necessary to support the forearm during flexion and extension. In most cases, the application of ice facilitates the exercises.

If the arm shows a tendency to swell, elevation and exercise in changing positions might be necessary. Starting with the fourth week the patient wears the cast only at night. The daily exercises now include pronation and supination, together with more complex movements. Hydrotherapy and swimming may also be added to the program.

After the healing is completed, seven weeks at the earliest, exercises against resistance with equipment and PNF can be started. Training in activities of daily living is included in physical and occupational therapy.

Sample Exercises

Depending on the phase, the following exercises may be performed:

- the first two exercises listed earlier for a surgically treated fracture of the humeral shaft
- exercises listed earlier for a nonsurgically treated supracondylar fracture of the humerus

Galeazzi Fracture

Symptoms and Therapeutic Aspects

The symptoms and therapeutic procedures for a Galeazzi fracture are the same as for a conservatively treated supracondylar fracture of the humerus with the addition of treatment for scar tissue.

Course of Treatment

On the first day, the patient is instructed to exercise according to the following previously described guidelines for prophylactic treatment of patients with injuries of the shoulder joint. The normal arm should support the injured arm during the shoulder exercises.

Beginning on the second day, the patient is allowed to be out of bed. The exercise program now contains edema prophylaxis, exercises in changing positions, and dynamic exercises of the hand and elbow. Pronation and supination are not yet allowed.

After the wound has healed, the local blood supply is improved by heat or cold modalities. Exercises in the therapeutic pool facilitate the mobilizing and strengthening therapy. After the fourth week, pronation, supination, and swimming are allowed.

When the consolidation of the fracture is completed (after about the seventh week), the patient exercises against resistance and participates in occupational therapy.

Sample Exercises

Depending on the phase, the following exercises are performed:

- the first two exercises listed earlier for surgically treated fracture of the humeral shaft
- exercises listed earlier for a nonsurgically treated supracondylar fracture of the humerus

DISTAL FRACTURES OF THE FOREARM

Radial fractures close to the wrist are very common.

Mechanism of Injury

The major cause of a distal fracture of the forearm is a fall on the hand in wrist extension (Colles' fracture) (see Figure 13–3) in which compression and bending forces are involved. A fall on the hand in volar flexion results in the opposite type of fracture (Smith's fracture).

Types

Usually, distal fractures of the forearm are compressed and dislocated, and occur in combination with a fracture of the styloid process of the ulna. The so-called bayonet position (radial and dorsal angulation) is characteristic. A subluxation of the radioulnar joint might occur together with a radial fracture in the typical location. A subluxation by itself is rare.

Clinical Considerations and Diagnosis

Malalignment, swelling, pain due to movement or palpation, and a limited range of motion are characteristic symptoms. Radiographic examination confirms the diagnosis.

Medical Management

The treatment of an acute fracture of the distal forearm is directed at an exact alignment under traction and local anesthesia. Afterward, a cast is applied for about 4 to 6 weeks. Unstable fragments can be secured by crossed Kirschner wires. Serial radiographs are taken after the first, fourth, and eighth days and after four weeks. Only in exceptional cases are an open reduction and osteosynthesis re-

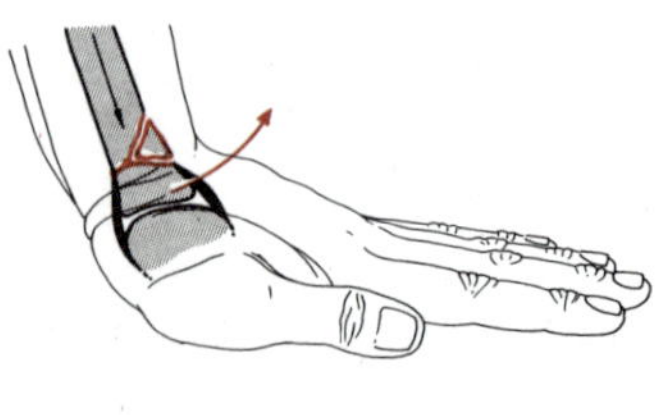

Figure 13–3 Distal fracture of the forearm that occurs when the hand is extended dorsally

Source: Schlosser and Kuner, 1980

quired. The simultaneous distal fracture of the ulna does not require special treatment, since it is realigned during the reduction of the radial fracture.

Early and Late Complications

The lesion of the median nerve is possible in a severe dislocation. A consolidation in malalignment or hypertrophic callus formation can cause a secondary carpal tunnel syndrome and a rupture of the extendor tendon of the thumb. In older patients, if there is a difference in the length of the radius and ulna as a result of the fracture and, consequently, post-traumatic joint degeneration, the ulnar portion close to the joint is usually resected. In young patients, an osteotomy is indicated to shorten the ulna or to straighten the radius. Frequent corrections and long immobilization periods can result in Sudeck's atrophy.

In infants and adolescents, radial fractures with typical symptoms can affect the epiphyseal growth plate. The traumatic lysis of the radial epiphysis is occasionally combined with a separation fracture of a metaphyseal bone fragment (Aitkens).

Physical Therapy

Symptoms

In both surgical and nonsurgical treatments, the fracture is immobilized in a cast and elevated on a wedge pillow. After removal of the cast, the following symptoms are observable:

- dry, scaly skin
- residual hematoma
- swelling around the wrist, sometimes also in the hand and fingers
- in many cases, a malalignment
- edematous tissue
- limited range of motion of the wrist
- atrophic and weak muscles
- pain

Therapeutic Aspects

The therapeutic procedures are basically the same as for nonsurgically treated supracondylar fracture of the humerus.

Course of Treatment

From the first to about the sixth week, if the radial fracture has not achieved stability needed for resistive exercises, nonsurgical therapy continues to be applied. The forearm is immobilized in a cast, and the patient is encouraged to exercise following previously described guidelines for the prophylactic treatment of patients with injuries of the shoulder joint. A correct fit of the cast must be ensured, and any swelling or discoloration must be carefully monitored to avoid Sudeck's syndrome.

Beginning in about the seventh week, the cast is worn only at night. In case of swelling, the hand and arm must be wrapped in an elastic bandage. Elevation and position changes precede the mobilization and strengthening treatment. The therapist exercises with the patient in free movement of the wrist. Only in exceptional cases are pronation and supination delayed. The exercises consist of active muscle work, the use of PNF techniques, and exercises with equipment: a ball, dumbbells, cane, and so on. Exercises at a specially equipped wall can be very helpful for retraining in daily activities and skills (see Figure 13–4).

To supplement the physical therapy, whirlpools, kneading theraputty, and occupational therapy are prescribed. The degree of weightbearing allowed is determined by the results of radiographic examination.

Sample Exercises

In the early phase of mobilization and strengthening, the patient sits at the treatment table and exercises the finger and wrist, if possible including supination and pronation.

Figure 13–4 Exercise wall

In the late phase of mobilization and strengthening, the patient uses PNF exercises in all diagonals, with fixation of the proximal joints and stretch to the hand and fingers. In the case of limited wrist extension, the patient also does crawling exercises.

In the late phase of mobilization and strengthening, the patient does weight-bearing activities and exercises in changing positions on an air pillow. Then, while in a kneeling position in front of a gymnastic ball, supported bilaterally with both hands, the patient rolls the ball forward (wrist flexion) and backward (wrist extension), stabilizing it at various points.

Finally, the patient does training exercises in daily activities, using the exercise wall (see Figure 13–4).

INJURIES OF THE MEDIAN NERVE

Mechanism of Injury

Paralysis of the muscles innervated by the median nerve and sensory deficits are caused by fractures of the distal humerus, by subluxations of the elbow, or by lesions close to the wrist.

Types

Frequently, the median nerve is compressed during subluxations or by fracture fragments or external pressure on the upper arm. This may also occur in subluxation fractures of the carpal bones. Cuts through the wrist may cause an interruption of continuity.

Clinical Considerations

The paralysis of the upper median nerve results in the typical monkey hand. Paralysis of the medial median nerve (middle of the forearm) or of the lower median nerve (wrist) is characterized by atrophy of the thumb muscles. Various kinds of sensory deficits result from abnormal innervations between the median and ulnar nerves. Typical deficits are seen in hypoesthesia of the middle and end phalanges of the dorsal side of the second and third finger, of the volar side of the thumb, second and third finger, and of the radial part of the fourth finger and the palm of the hand proximal to it.

A special complication is the carpal tunnel syndrome. In this case, the median nerve is compressed below the retinaculum flexorum and the transverse carpal ligament. The syndrome usually appears after an old fracture of the wrist or as a result of degeneration of the carpal bones. More frequently, it results from thickening of the tendon sheath. Typical symptoms are pain in the hand and fingers during the night, paralysis and atrophy of the thenar eminence. A pain response can be initiated by compression of the median nerve.

Medical Management

The treatment consists of exposing the nerve at the site of the lesion, decompression, and, if there is a laceration, suture. An arthrodesis of the carpometacarpal joint is indicated for atrophic muscles of the thenar eminence.

In the carpal tunnel syndrome, the transverse ligament of the carpal ligament is split and excised. Combined lesions of the ulnar and median nerves can occur because of the close proximity of the nerves along the inside of the upper arm, forearm, and wrist.

Physical Therapy

A paralysis of the median nerve should be treated with the following procedures:

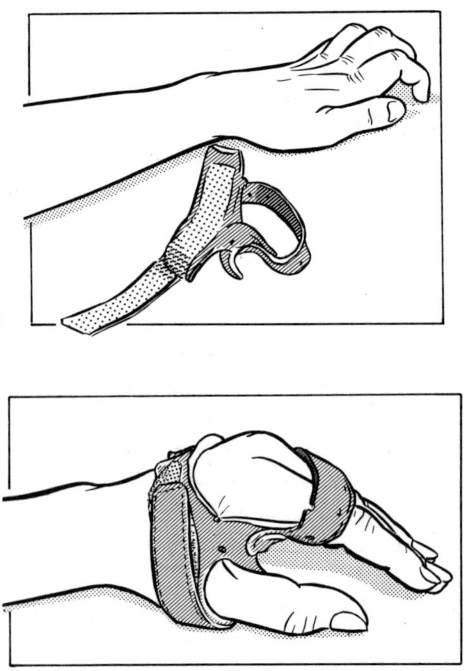

a

b

Figure 13–5 Positioning splint for paresis of both the medialis and ulnaris nerves

- support and prevention of a contracture by a splint (see Figure 13–5)
- improvement of the blood supply by active exercises
- electrotherapy, provided there is no implant material that might interfere with the electrical field
- stimulation by ice applications
- exercises to support innervation and for strengthening, using PNF techniques as much as possible
- training in activities of daily living with physical and occupational therapy

Chapter 14

Injuries of the Wrist and Hand

Injuries of the hand and their treatment have a particular clinical significance. The integrity of the hand for touching, catching, and grasping in daily activities is essential. The treatment of the hand thus must take into account all of its complex functions. This has required the development of special techniques by both physicians and physical therapists. The specialized literature should be consulted for detailed information on the continually developing techniques used in hand surgery.

One of the important basic techniques in the surgical treatment of hand injuries is careful manipulation. This technique is also basic in the application of physical therapy for such injuries.

For open wounds and in surgical treatment of closed injuries of the hand, careful control of hemorrhaging and surgical debridement with appropriate instruments surrounding damaged tissue are mandatory. In the process of functional restitution, the treatment should ensure that opposition of the thumb is maintained and that both sensory and motor functions are maintained or restored. The postoperative physical therapy is also important in the restitution of a functional hand; it must be started as early as possible and must be applied carefully.

The injuries of the hand are frequently multiple injuries that take various forms. If there is extensive fresh soft-tissue damage, especially with pending amputation or partial resections, therapeutic procedures must be delayed.

The wound is first cleaned (irrigation, debridement) and a sterile bandage with aseptic dressing is applied. During the subsequent healing period, vital tissue will recover and tissue that has been severely damaged will start necrotizing. This necrotic tissue should be debrided. Normally, the objective is to obtain a skin coverage with minimal scarring. All other structures of the hand can be restored secondarily.

In a case of multiple trauma, skin and tendon injuries are treated and fractures are stabilized immediately. The treatment of the injured flexor tendons and nerves is delayed.

INJURIES OF THE SKIN

Mechanism of Injury

Both sharp and dull objects can cause cuts and contusions. Subcutaneous structures, such as tendons, blood vessels, nerves, bones, and joints, can also be involved.

Medical Management

A fresh clean wound should be closed without any tissue tension. Bite wounds (animals, humans) and bullet wounds are treated openly. More extensive skin lesions have to be covered with transplants: split-thickness skin grafts on the extension side, full-thickness skin or flaps on the flexor side. Amputations of fingers must be done with great caution in order to preserve the sensory and motor function of the finger. Maintenance of the opposition function of the thumb is especially important, justifying all attempts to save, if necessary by resuturing, a completely severed thumb.

INJURIES OF THE TENDONS

The tendinous tissue heals slowly and has a tendency to result in adhesions and inflammation. The treatment of flexor tendons and extensor tendons is quite different. A completely severed tendon in an open clean wound is sutured end-to-end, either primarily or secondarily. Closed ruptures or separations from the bony insertion are immobilized for 4 to 6 weeks in extension. Because of their special anatomical structure, the flexor tendons within the tendon sheaths are frequently not repaired during the primary treatment of the injury. If the flexor tendon is ruptured outside the sheath, the tendon is sutured primarily. Usually, in lesions of both the superficial and deep tendon, only the deep tendon is repaired. After a wound has healed completely, a secondary transplant with the long palmar tendon to replace the function of the superficial tendon is necessary. Active physical therapy is started after an immobilization of 3 to 4 weeks.

SUBLUXATIONS

Subluxations of the wrist, metacarpal, and finger joints are reduced immediately, usually by traction. If the attempt to achieve a closed reduction fails or portions of the flexor tendons or joint capsules (first finger joint) are trapped, an open reduction is required. The open reduction is combined with a suture of the ligament and capsule. This procedure is also required in subluxations of the middle finger joints (disruption of the volar capsule and subluxation to the flexor side). In the following sections, the most common subluxations and fractures and particularly significant injuries of the tendons are discussed.

INJURIES OF THE NERVES

In a clean wound, a severed nerve is sutured primarily if the separated ends are smooth and the skin can be closed without tension. In a contaminated wound, the nerve is repaired after the wound heals. Defects have to be bridged.

WRIST SUBLUXATION

Mechanism of Injury

Subluxations of the wrist are caused by a fall on the hand. However, they occur far less frequently than radial fractures caused by similar falls because of the wrist's strong, ligamentous apparatus.

Types

The most frequent type of wrist subluxation is a subluxation of the lunar bone (see Figure 14–1). The lunate bone remains in place and the wrist subluxates dorsally. This injury can occur in combination with a subluxation fracture of the scaphoid in the form of a transscaphoid perilunate subluxation. In isolated semilunate subluxations to the volar side, the bone turns 90 degrees. These subluxations might be missed initially. The symptoms of a perilunate subluxation are swelling, pain, a limited range of motion, and a malalignment with bony malformation (visible or palpable). Frequently, the median nerve is compressed.

In the case of a fresh acute trauma, the repositioning can be accomplished using muscle relaxants and extension of the wrist. Subsequently, surgical reduction of

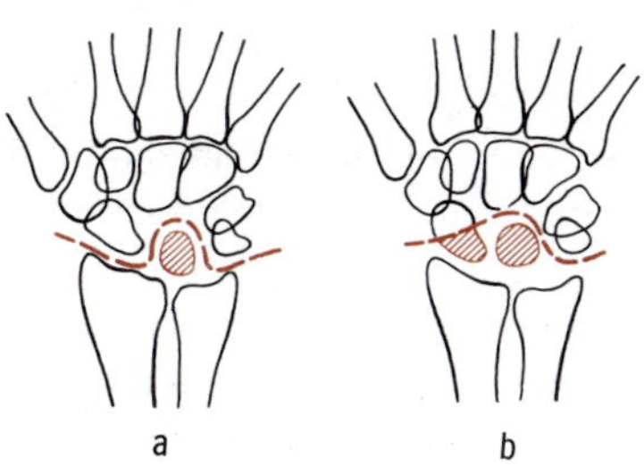

Figure 14–1 Subluxations of the wrist: a. perilunate subluxation; b. de Quervain fracture

Source: Schlosser and Kuner, 1980

the lunate bone is indicated. Afterward, the reduced subluxation is immobilized in a cast for 6 to 8 weeks. If the scaphoid is fractured at the same time, it has to be fixed by screws. In this case, the immobilization period in a cast is 10 to 12 weeks. This is followed by physical therapy and training of the hand in occupational therapy.

WRIST FRACTURES

The fracture of the scaphoid bone is the most frequent fracture of the wrist.

Mechanism of Injury

A fracture of the scaphoid bone is caused by a fall on the extended, ulnarly deviated hand. It can also be caused by a direct impact on the radial portion of the wrist.

Types

Fractures of the wrist joint include transverse fractures of the radial, median, and ulnar portions.

Clinical Considerations and Diagnosis

The symptoms are swelling (frequently only mild) around the wrist, local pain, and sensitivity of the thumb to compression, traction, and contusion. The diagnosis must be confirmed by special radiographs. Since some early fissures can be missed, subsequent radiographs after 10 days are recommended; the fissure will then be more visible by bone resorption.

Medical Management

Treatment consists of complete immobilization in a cast for three months. The immobilization includes the thumb and the first phalanges of the second and third fingers and, during the first four weeks, also the upper arm. Dislocated fragments can also be fixed by screws, which shortens the immobilization period.

Except for separation fractures of the cuneiform bone, fractures of the carpal bones are rare. Fractures of the lunate bone are believed to cause necrosis of the lunate bone; however, such necrosis is more likely to be caused by disposition and microtrauma, which can occur while working with a pneumatic hammer. All other bone lesions of the wrist are usually simultaneous injuries in a multiple trauma of the hand. The treatment is focused on the main injury.

Early and Late Complications

Delayed bony healing and pseudarthrosis are common in proximal fractures, because of the radially located blood supply, and require surgery. Various specialized techniques are used, in some cases together with electrostimulation (Zichner, 1979). This treatment is followed by an immobilization period of several months. Afterward, intensive physical therapy is necessary.

METACARPAL FRACTURES

Bennett Subluxation Fractures

A Bennett fracture is a dislocation fracture of the first metacarpal bone at the carpal-metacarpal joint. Among the fractures of the metacarpals, this fracture has special significance, since the principal movements of the thumb involve the first metacarpal bone and the trapezoid bone. The first joint of the thumb is a hinge joint and allows movements only in one plane.

Mechanism of Injury

The Bennett fracture is caused mainly by a compression of the adducted thumb along the longitudinal axis.

Types

The Bennett fracture is an oblique fracture of the base of the first metacarpal bone, extending into the saddle joint. The small medial fragment remains connected to the joint, while the first metacarpal bone is shifted into dorsal, radial, and proximal directions. Besides this characteristic fracture, extra-articular fractures at the first carpal-metacarpal joint and Y- and T-shaped fractures (Rolando fractures) can occur.

Clinical Considerations and Diagnosis

The symptoms are pain, swelling, and malalignment. Radiographs are used to reveal the severity of the lesion.

Medical Management

Treatment is aimed at reduction by traction on the abducted thumb and fixation of the thumb in a cast. Osteosynthesis with Kirschner wires and a cast in abduction or with screw fixation of the fragments may also be necessary.

Metacarpal Fractures

Mechanism of Injury

Fractures of the metacarpals are caused by contusion or bending due to a fall on the hand. They can also result from a strong direct impact.

Types

The fractures are located at the base, at the shaft, or close to the head of the metacarpal bone. They vary from fissures to transverse, oblique, and rotational fractures to compound and comminuted fractures, with or without dislocation. They can also be combined with subluxations in the carpal or carpometacarpal area and lesions of ligaments and capsules. Comminuted fractures may be accompanied by extensive lesions of the soft tissue, especially tendons and tendon sheaths and nerves. Their treatment has absolute priority in a multiple trauma.

Clinical Considerations and Diagnosis

The symptoms of a fracture of the metacarpal are pain, swelling, a hump on the dorsum of the hand, palpable ridges, and abnormal mobility. Radiographs are used to reveal the extent of the injury.

Medical Management

The treatment is usually nonsurgical. After the alignment of the fracture, the hand is immobilized in a functional position for 1 to 2 months. In multiple fractures, stabilization by osteosynthesis of the second and fifth metacarpal bone is preferred. If, in fractures of the shaft and fractures in close proximity to the metacarpal heads, there is insufficient reduction or strong deviation, surgical reduction and fixation are required. Stable osteosyntheses allow early active exercise therapy. The materials used in the osteosynthesis are either percutaneously inserted Kirschner wires or, in an open fixation, implanted plates and screws. In case of extensive tissue damage, the stabilization is done with Kirschner wire and is then completed by a cast immobilization for 3 to 4 weeks. The essential concern is restoration of the best possible functional ability.

Complications

The most serious complication is infection. An incomplete reduction can cause a shortening and a rotary malalignment, which can seriously restrict the function of the hand. In such cases, corrective surgery might be necessary.

PHALANGES FRACTURES

Mechanism of Injury

Fractures of the phalanges are caused by compression or torsion.

Types

Fractures of the phalanges usually occur at the base and close to the head, sometimes also intra-articularly. The proximal phalanges are most often affected.

Clinical Considerations and Diagnosis

The symptoms are malalignment, bony malformation, and abnormal mobility. The diagnosis is confirmed by radiographs.

Medical Management

The treatment of a simple fracture of the phalange is limited to nonsurgical reduction under traction and immobilization of the fracture for four weeks, with the base joint in 90 degree flexion and with an approximately 45 degree flexion of the remaining joints (if necessary, including the next finger). If the nonsurgical treatment is not successful, the fracture is corrected surgically (fixation with Kirschner wires) with subsequent immobilization for four weeks. Fractures of the distal phalanges do not require special treatment. Fractures of bony ligament insertions and fractures with bony malformation within the joint must be reduced and fixed with screws or wires.

INJURIES OF THE EXTENSOR TENDONS

Mechanism of Injury

Injuries of the extensor tendons occur frequently in multiple trauma and in trauma resulting from daily routines. They may include stab wounds or lacerations.

Clinical Considerations

Loss of active extension is the characteristic symptom.

Medical Management

The treatment depends on the clinical requirements of the multiple trauma. Isolated ruptures are repaired by sutures. An acute rupture at the finger joint can be treated nonsurgically by immobilization in extension for approximately six weeks. Otherwise, particularly in chronic ruptures, the tendon must be tightened by a suture. A separation of the bony insertion (subluxation fracture) is surgically realigned and fixed. Surgery is also required in case of a rupture of the transverse or longitudinal medial portion ("buttonhole" rupture) of the extensor aponeurosis across the middle joint. A transition of the tendon is recommended for ruptures of the long extensor tendon of the thumb due to degenerative changes, bony malformation, and so on.

INJURIES OF THE FLEXOR TENDON

Mechanism of Injury

The flexor tendon can be injured in a variety of ways, including open cut and stab wounds.

Clinical Considerations

The characteristic symptom is the inability to bend the fingers actively.

Medical Management

The accompanying lesions of bones, vessels, or nerves have to be considered. The sheaths of the flexor tendon have a tendency to develop adhesions, which frequently inhibit suture repair. Therefore a secondary treatment, possibly with a transplant, is recommended. A primary suture repair is limited to fresh uncomplicated wounds.

PHYSICAL THERAPY FOR INJURIES OF THE HAND

The increasing number of hand injuries and unique importance of the hand as a highly specialized complex tool underscore the risks of erroneous treatment and have led increasingly to the establishment of facilities to deal exclusively with hand injuries. Specialized training is mandatory for physicians involved in the treatment of the hand. Such training should also be required of the physical therapist, especially in the early phases of the treatment.

It is not feasible to discuss in detail the great variety of injuries of the hand and their physical therapies. In the following pages, only the basic principles of treatment are presented. Specialized literature should be consulted for further specific information on injuries of the hand and their treatment.

Conditions for an Optimal Treatment

The following are prerequisites for optimal treatment of hand injuries:

- knowledge of the anatomy and the function of the hand
- knowledge of the initial medical treatment and its relevance for subsequent therapy
- precise information from the physician/surgeon concerning the required therapy
- ability to evaluate the patient and the injury precisely and thoroughly
- ability to select the correct therapeutic program to meet the patient's needs
- knowledge of the relevant special techniques
- communication between the physician and therapist during the therapy
- communication and cooperation with those involved in parallel therapies, especially occupational therapy
- motivation by and cooperation from the patient

Criteria for Selecting Therapeutic Procedures

The following factors should govern the choice of treatment and the way in which it is applied in the individual case:

- type of injury
- type of treatment (techniques of surgical or nonsurgical treatment)
- result of the medical treatment
- results of examination by the physical therapist
- reactions to the treatment in the particular case

Therapeutic Principles

Even a minor trauma can cause complications in the function of the hand with its differentiated structures. Indeed, more than any other part of the body, the hand is unforgiving of overly intensive manipulations that result in severe irritations. Thus, it is essential that the following principles be recognized in working with injured hands:

- The physician is responsible for immobilizing the injured arm correctly, the therapist for elevating it properly, and for instructing the patient to exercise following guidelines for prophylactic therapy of patients with injured arms.
- Active movements of the hand should begin as soon as possible.
- The therapist should ensure pain-free positions and careful support and should respect the patient's pain threshold during the treatment.
- Normally, during the early phase of the treatment, passive treatment and direct application of heat should be avoided.
- During the late phase of the treatment, passive mobilization and the application of heat should be done with extreme caution.
- The patient should be motivated to do independent exercises as soon as this is possible without risk.

The following detailed information is required as a basis for physical therapy:

1. personal data
    - name
    - age
    - occupation
    - date of accident
    - diagnosis
    - surgical report
2. basic information
    - vital statistics
    - case history
    - radiographs
3. visual observation
    - scar
    - discoloration
    - edema
    - alignment position of the bone components within the joint
    - use, function

4. tactile observation

    - sensitivity
    - sweat secretion

5. measurements

    - range of motion of the joints
    - circumference

6. evaluation of function

    - muscle test
    - forms of grasp and pinch: key, chuck (three-point), tip, etc.
    - active movements

7. supplementary observations

    - readiness to cooperate
    - social background

Therapeutic Goal

The goal of each treatment is to achieve the best possible restitution of the function of the hand. To this end, the following therapeutic aspects must be taken into account:

1. relaxation

    - painless position
    - applications of ice
    - general and local relaxation exercises

2. edema prophylaxis

    - elevated position
    - isometric and dynamic exercises
    - effleurage massage

3. support of local blood supply

    - isometric and dynamic exercises
    - applications of ice
    - hydrotherapy
    - kneading theraputty (initially 38°C, subsequently 40°C, if tolerated)

4. mobilization

    - splint (at night)
    - treatment of the scar, massages, special ointments
    - active movements
    - relaxation techniques
    - manual therapy
    - in case of tendon injuries, careful passive movements and stretching, wrapping the fist

5. muscle facilitation techniques

    - cryotherapy
    - stimulating exercise techniques
    - electrotherapy

6. training in activities of daily living

    - active exercises
    - exercising grasp and pinch forms: key, chuck (three-point), tip
    - training in functional skills

7. strengthening

    - exercises against resistance
    - exercises with equipment

Simultaneous Therapeutic Possibilities

Occupational therapy should, if possible, be applied simultaneously with physical therapy.

A Case Example: Fracture of the Scaphoid Bone

The following sections describe the process by which the therapist treats a fracture of the scaphoid bone. The procedures discussed constitute, in effect, a model treatment that can be applied to the lesions of bone structures presented in the general medical part of this book. In each case, it is necessary to modify the therapeutic procedures and program to meet the specific needs of the individual situation.

Basic Medical Treatment

1. nonsurgical
   - dorsal splint (including upper arm) for three days
   - after the swelling subsides, a circular cast from the upper arm to the first metacarpal bone, leaving the fingers free beyond the metacarpophalangeal joints
   - immobilization period of at least six, sometimes up to 12 to 14 weeks, with the cast around the upper arm replaced after six weeks by a forearm cast
2. surgical
   - osteosynthesis with compression screw
   - dorsal cast (including upper arm) for about 10 days
   - after the wound has healed, a circular cast, including the upper arm
   - immobilization period of approximately six weeks
3. position in both surgical and nonsurgical treatments: during the first 3–5 days, on an abduction pillow, hooked to a loop (see Figure 12–13).

Physical Therapy

In this case, the treatment is the same for surgical and nonsurgical treatments.

Symptoms

1. visual observation
   - swelling around hand and fingers
   - possible discoloration
   - atrophy of muscles of the hand, sometimes the arm

2. tactile observation

   - possibly slightly elevated local temperature
   - postoperatively, severe scar adhesions
   - diminished muscle tone, but increased muscle tightness
   - hypersensitivity due to inactivity
   - pain when attempting to move

3. general function

   - limited range of motion, especially of the wrist, but sometimes also of the elbow, fingers, and thumb
   - limited function of grasping and pinching
   - tendency to remain inactive

Therapeutic Goal

The therapeutic goal is to achieve restitution of a complete and painless function.

Therapeutic Program

1. relaxation

   - pain-free position
   - general and local relaxation exercises
   - cryotherapy

2. edema prophylaxis

   - positioning
   - isometrics and dynamic exercises

3. improving local blood supply

   - isometrics and dynamic exercises
   - applications of ice
   - hydrotherapy
   - kneading theraputty

4. mobilization

   - scar care (creams, massage)
   - active movement of the entire arm
   - active movement of the hand
   - active exercises
   - relaxation techniques

5. training in activities of daily living

   - exercising grasp and pinch
   - active movement
   - training in functional skills

6. strengthening

   - exercises against resistance (PNF, isolated muscle contractions)
   - exercises with equipment

Course of Treatment

During the immobilization period in the cast, the physical therapist, along with all others involved in the treatment, must ensure that the cast is positioned correctly. The therapist must also instruct the patient in exercises following guidelines for prophylactic treatment of patients with injured arms.

After the removal of the cast and subsequent skin and scar care, the therapy continues with relaxation exercise and edema prophylaxis. If the patient is apprehensive and anxious, the exercises for relaxation can serve as a useful introduction. Training in relaxation can also be done with PNF techniques. During this early phase of the treatment, the physical therapist should work only with the muscles proximal to the fracture.

After the removal of the cast, the hand and fingers frequently have a tendency to swell. These conditions can be effectively limited by elevation and applications of ice (stroking from distal to proximal) and by active muscle contraction. Hydrotherapy and intensive active movement help to improve the local blood supply.

Kneading exercises with warm theraputty (38°C) have a mildly stimulating effect on the local circulation. They also facilitate mobilization and strengthening, and most patients enjoy doing them. Mobilizing exercises are scheduled not only for fingers and hand but also for the elbow and shoulder. The exercises for the fingers and hand are, as always in a fresh injury of the hand, strictly active. They can include a group of muscles or they can be worked selectively, following relaxation techniques. The newly achieved motion is maintained and strengthened by repeated contractions.

Exercises in the various grasp and pinch forms are essential (see previous sections on symptoms and therapeutic aspects in injuries of the hand). These exercises are modified in the light of the patient's occupation or requirements.

Sample Exercises

The following exercises should be performed in the middle and late phases of mobilization, training in activities of daily living, and strengthening:

1. PNF: exercising in (1) a bilateral symmetrical arm pattern in a supine position, (2) in a combination of shoulder-arm in all variations, and (3) in both diagonals employing all possibilities and directions and including the elbow; also using the overflow in the action of the hand and fingers by stabilizing the strong part
2. exercises in a supine or sitting position, with the forearm on the treatment table: (1) finger and hand flexion in combination with flexion-supination of the elbow, then (2) finger and hand extension in combination with extension-pronation of the elbow; doing these exercises first with a leading contact, then with repeated contractions in a slow return, slow return-holding sequence
3. exercises in a sitting position at the treatment table:
    - selective exercises for wrist and finger joints, first only actively, then with leading contacts and resistances
    - exercising forms of grasp and pinch, first only actively and then with resistances
4. exercising with various types of equipment
    - digging in a pot filled with peas, beans, rice
    - kneading theraputty: rolling, forming, pulling
    - turning the lids of jars of different sizes
    - moving a rolling pin
    - drying ink with a blotter
    - cutting and crumpling paper
    - stirring ice cubes in a bucket
    - threading a needle

- picking up objects
- buttoning and unbuttoning a shirt

5. weightbearing on the hand

- on an air pillow (see Figure 9–3)
- on a gymnastics ball
- on a thick floor mat
- on a hard floor

Physical Therapy for Injury of the Extensor Tendons of the Finger

During the postoperative immobilization period, whose length depends on the level of disruption, the type of injury, rupture, disruption of the insertion, and so on, and the surgical technique employed, the patient is encouraged to exercise following guidelines for prophylactic therapy in patients with injured arms. After the immobilization period, the patient is treated according to the following program:

1. edema prophylaxis and improving local blood supply

- possibly elevation
- isometric and dynamic exercises, sometimes with ice applications
- active movements in warm water
- kneading theraputty

2. mobilization

- scar treatment
- active exercises (selective extension, especially the metacarpophalangeal joint, increasing flexion, moving from flexion to extension)
- active movements in warm water
- kneading theraputty

3. training in daily function and routines

- active movements
- exercising forms of grasp and pinch
- active work with various types of small everyday objects and devices

4. strengthening: active movement against resistance and with various types of equipment

The patient is encouraged to participate simultaneously in occupational therapy.

Physical Therapy for Injury of the Flexor Tendon of the Finger (Kleinert Suture)

After the Kleinert suture, the patient is equipped with a dynamic splint (see Figure 14–2). As usual, the physical therapy is started the day after the operation.

Initially, the injured finger is held passively in flexion by the rubber band (relieving the suture). As a basic exercise, the therapist works with the patient in active extension of the finger against the splint. It is important that the patient reach the full extension of the middle joint, which has the greatest tendency to result in a contracture. The finger is led back passively into flexion by the elastic rubber band (no active movement). The patient is instructed to do this exercise independently several times each hour, at the same time, moving the elbow and shoulder joint.

Course of Treatment

After about three weeks, the dynamic splint is replaced by an elastic bandage around the wrist, with the rubber band hooked to it. This prevents overuse of the patient's finger and hand, which could again rupture the tendon.

The therapist places the wrist in flexion, supported by a sandbag, and works selectively on an active flexion and extension (the latter with particular care) of the metacarpophalangeal joint and the proximal and distal interphalangeal joint.

After about five weeks, the elastic bandage and the rubber band can be removed, and the selective exercises can be intensified. The scar is treated by a gently rotating massage and special ointment; this treatment contributes greatly to the functional improvement.

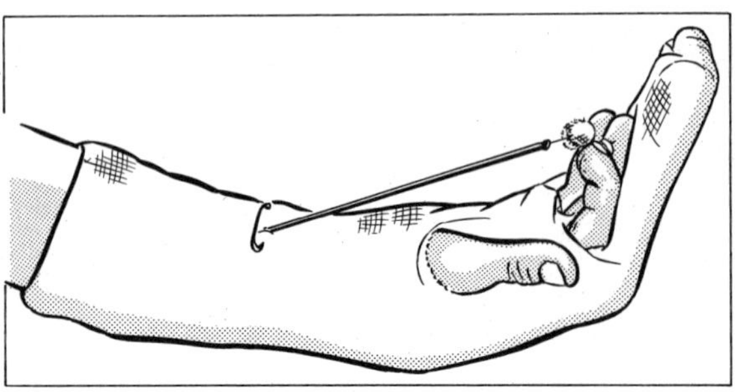

Figure 14–2 Dynamic splint after a Kleinert suture

After about six or seven weeks, the therapist starts to supplement the therapeutic program by carefully increasing to complex exercises and training in activities of daily living. In addition, the patient kneads theraputty and participates in occupational therapy. Possible flexion contractures are treated by gentle, passive stretching into extension. In occupational therapy, the patient is equipped with a splint.

The specialized literature on surgery of the hand should be consulted for additional details regarding treatment.

Chapter 15

Disturbances in Healing

SUDECK'S ATROPHY (REFLEX SYMPATHETIC DYSTROPHY)

Local overactivity of the sympathetic nervous system, as described by Sudeck, may be the cause of the vasomotor changes and pain responses following a disease or trauma.

Mechanism of Injury

Contributing factors are the trauma itself, inadequate care of the tissue during the treatment, repeated attempts to correct the bony alignment, and constricting splints or casts. Partial lesions of the nerves are also considered to be a causal factor; these lesions result in increased permeability of the cell and the cell membrane.

Types

Sudeck's atrophy is observed more frequently in the upper than in the lower extremities. Three stages are distinguishable: acute, subacute, and chronic.

Clinical Considerations and Diagnosis

Patients with Sudeck's atrophy usually have more intense pain response than one might expect with the extent of trauma. Degenerative changes of the cervical spine might also contribute to the syndrome. A suspected psychological instability of a patient might lead one to assume that previous experiences related to illness and hypersensitivity might be involved. Considering all of these factors, treatment in the early stages of a Sudeck's atrophy is necessary for a complete healing.

Therapeutic Stages

The early stage of the atrophy occurs approximately four weeks after the injury. An early symptom is a burning pain that recurs spontaneously in the joints and muscles and to a lesser extent in the bones. The pain is accompanied by hyperemia, exudation, and proliferation. This first stage is characterized by elevated temperature and discoloration of the skin; the wrinkles disappear, and the skin looks shiny. The pain occurs spontaneously but increases with exercise. At this stage, radiographs will not reveal any characteristic symptoms.

Medical management in the first stage, if not already done, is immobilization to eliminate any irritations. It is also important to make sure that the blood circulation is not disturbed by bandages or a cast. Active exercises for relaxation and stimulation are prescribed, together with ice applications and correct positioning. The drug therapy might include cortisone, calcium, sedatives, or anti-inflammatory agents.

In the second stage (after 2 to 4 months), the characteristics of a chronic inflammation with dystrophy appear: diminished blood circulation and metabolism, shiny skin, changes of the nails, and cyanosis. The spontaneous pain is replaced by pain during moving and weightbearing of the partially or completely limited extremity. Finally, the affected extremity turns blue or gray, the skin feels cool, and the secretion of sweat and oil is reduced.

The radiographic examination now reveals bone resorption, especially of the spongy bone; patchy areas of decalcification close to the joint; and, finally, thinning of the cortex. The muscles have lost their tone and the joints have a very limited range of motion.

Medical management in the second stage: A fairly complete restitution is still possible. Medical treatment must be based on clinical observations, which frequently show improvement while the radiographic results still remain negative. The mobilizations by physical therapists must be applied very carefully; the exercise requirements might have to be reduced to avoid moving into the third stage. Movement, applications of ice, and passive stretching are particularly important. Drugs are no longer effective.

In the third and final stage, the atrophy can no longer be reversed completely. The skin is now blue and cyanotic with abnormal growth of hair and nails and abnormal sweat secretion. The joints are stiff because of capsular contractions; the muscles are atrophic, and the pain is decreasing. Radiographs reveal a coarse structure of the spongy bone and a thin cortex.

In addition to splints and slings, therapy in the third stage is applied in the form of passive and active exercises. Frequently, occupational retraining is necessary.

Vigorous manipulations should be avoided. The treatment should be carefully applied, with early active exercises of the nonimmobilized joints, including the noninvolved side. Constrictive bandages should be avoided. In the case of anxious patients, the condition might be improved by application of calcium and sedatives.

Physical Therapy

The atrophic syndrome can develop in either the upper or lower extremity. However, since the most severe complication of the healing process occurs mostly in the upper extremity, the following discussion is concerned only with treatment of that extremity. The same therapeutic principles are, with only minor modifications, also applicable to the lower extremity.

Prophylaxis

The therapist explains to the patient the importance of the prophylactic therapeutic procedures described in the first part of the text. The therapist works with the patient accordingly and instructs the patient in exercising independently. The therapist should at all times ensure the correct fit of casts and bandages and be aware if instruction in movement and positioning is needed.

Physical Therapy

In each phase of the treatment of patients with the atrophic syndrome, the therapist must apply the treatment procedures with great care, watching for any signs of incompatibility or stress, increased swelling, hyperthermia, and pain. If such signs are observed, the initiating factor has to be found and the program must be reevaluated and modified.

Active exercises are always preferable. However, if passive exercise is necessary, as is sometimes the case in the third stage, the therapist must proceed even more carefully.

First Stage

Symptoms

- redness
- swelling
- elevated local temperature
- increased sweat secretion
- limited range of motion, decreased or total loss of function

- spontaneous pain, frequently in combination with pain around the neck and shoulders
- increased muscle tightness

Therapeutic Aspects

1. relaxation

    - pain-free position
    - breathing and relaxation exercises

2. edema prophylaxis and improvement of local blood circulation

    - elevated position
    - possibly, careful isometric and dynamic exercises
    - cryotherapy (stroking with ice)
    - contrast baths (from 36°C to 18°C)

3. mobilization and strengthening: active movements (also PNF) with the normal arm and with the neck and shoulders

Second Stage

Symptoms

- skin discoloration
- shiny skin
- local hyperthermia
- muscle atrophy
- limited range of motion, partial or complete loss of function
- pain
- anxiety due to pain

Therapeutic Aspects

1. relaxation

    - pain-free position
    - breathing and relaxation exercises
    - PNF, proximal areas

2. edema prophylaxis and improvement of local blood circulation

    - elevated position
    - careful isometric and dynamic exercises
    - cryotherapy (stroking or exercises in ice water)
    - connective tissue massage

3. mobilization and strengthening

    - active movements, in combination with cryotherapy, with or without equipment
    - active movements (also PNF) with the normal arm and the neck and shoulder

4. training in activities of daily living: retraining of lost functions

Third Stage

Symptoms

- grayish pale, cool and dry skin
- sensitivity to cold temperature
- atrophy of the soft tissue
- contractures
- considerable loss of function
- pain

Therapeutic Aspects

1. possibly, edema prophylaxis

    - elevation, changing positions
    - isometric and dynamic exercises

2. stimulation of blood flow

    - warm baths, contrast baths
    - underwater massage avoiding all areas near the atrophic area
    - hydrotherapy

- connective tissue massage (local)
- kneading theraputty (38°C)
- isometric and dynamic exercise

3. mobilization

- active movements
- relaxation techniques, including mild passive stretching at the end of the motion
- manual therapy

4. strengthening and training in daily activities

- active movements, against resistance
- retraining in activities of daily living

DELAYED FRACTURE HEALING: PSEUDARTHROSIS

Mechanism of Injury

A delayed bone healing and the formation of a false joint are caused by insufficient contact among and immobilization of the bone fragments and interference with the blood supply. The delayed healing can be a direct result of the type of fracture or it can be due to shearing or twisting forces. It may also follow surgical treatment, because of interfering implant material, malalignment of the bone, poor blood supply, or infections.

Types

The term "delayed fracture healing" is used when a bony consolidation is not complete after a period of 6 to 9 months. After 9 to 12 months, the lack of bony consolidation is called "pseudarthrosis" (nonunion). The so-called hypertrophic pseudarthroses with hypertrophic callus formation due to instability are distinguishable from the so-called hypotrophic types, which are caused by the resorption of fragments and a resulting bone defect due to a lack of blood supply or to infection.

Clinical Considerations and Diagnosis

A typical pseudarthrosis is characterized by unstable fragments moving against each other. Their medullary cavity near the fracture site is closed and is covered histologically by a cartilaginous layer. Sometimes the pseudojoint capsule is filled with fluid. The symptoms of a pseudoarthrosis are an elastic, abnormal mobility, deformation, and pain, especially during weightbearing. Radiographic examination reveals the type and location of the fragments.

Medical Management

In the treatment of a pseudarthrosis, it is important to know whether it still has a vital, even if decreased, blood supply. Its ability to form callus is good and is even characterized by hypertrophic callus formation. A hypotrophic pseudarthrosis with insufficient blood supply usually has necrotic fragments and osseous defects and offers unfavorable conditions for fracture healing. As in an infected pseudarthrosis, its biological regenerative ability is lacking.

The medical management depends on the conditions. The treatment consists of a stable fixation by osteosynthesis. This induces the healing of the pseudarthroses. Hypotrophic pseudarthroses have to be filled with transplanted bone. Autologous spongy bone has the highest potency to form new bone and thus is used for bone transplants to replace bone fragments or to bridge defects. The combination of plate osteosynthesis and transplanted bone has proved to be successful in achieving and maintaining a stable osteosynthesis. In addition, electrostimulation and splints or bandages can be considered. Since a force along the longitudinal axis stimulates the healing of a pseudarthrosis, pseudarthroses in the middle of the shaft can be subjected to weightbearing with an intramedullary rod.

Physical Therapy

Surgical treatment is far more common than nonsurgical treatment. Physical therapy can be started as soon as exercise stability is achieved. The symptoms, therapeutic aspects, and treatment are basically the same as for fractures that heal in the normal period of time. The essential treatments are edema prophylaxis, improvement of blood supply, mobilization and strengthening, and, finally, training in daily activities. Exercises that exert shearing or lateral forces are not allowed.

In the case of a delayed fracture healing, as with normally healing fractures, the exercises are done without manual resistance until the consolidation of the fracture. However, it has been shown that the consolidation can be facilitated if pseudarthroses of the lower extremity are subjected to early, partial, axial weightbearing. The exact weight is determined by the physician. Generally, the patient at the same time begins exercises in the therapeutic pool, in deep water (up to the shoulders or chest) and with partial weightbearing (at least toe-touch weight) on crutches. To avoid unnecessary complications, it is important that the patient understand and carefully follow the therapist's instructions.

OSTEOMYELITIS

Mechanism of Injury

Traumatic osteomyelitis is an inflammatory consequence of the trauma and/or the treatment. The condition has significant implications for the success of the bone injury. It is caused by infectious microorganisms that invade the tissue during compound fractures or surgery. Correct wound care in open fractures and stabilization of the fragments afford the best prevention of osteomyelitis. An infection of the bone is rare in soft-tissue injuries.

Types

The main pathogenic agents are Staphylococcus aureus, Proteus organisms, and Escherichia coli. The location of these agents is dependent on the fracture site. In an osteosynthesis, it depends on the size of the implant. A special form of osteomyelitis that occurs after intramedullary nailing is the medullary inflammation.

Clinically, acute, subacute, and chronic osteomyelitis are distinguishable. The acute form becomes obvious only a few days or weeks after the injury or operation. The chronic form might not appear for many months or even years after the trauma or the implantation of osteosynthesic material.

Clinical Considerations and Diagnosis

The symptoms are pain and swelling, as a local sign of the inflammation, and fever, as a general reaction of the body. Later on, pus accumulates locally and the bone starts to necrotize until the formation of sequesters occurs.

Medical Management

The primary focus of infecton has to be cured; this usually requires surgical evacuation of the pus and excision of the infected tissue. If possible, stabilizing material is left in place and removed only in the case of a consolidated fracture. Necrotic fragments are removed, and defects are filled with autologous spongy bone, even in an infected pseudarthrosis. Rinsing with antibiotics or implantation of antibiotics have proved to be effective techniques. Coverage for the skin, possibly by skin grafts, is an important condition for healing. Late reoccurrence is possible.

Physical Therapy

Surgical Treatment

In recent years, surgical treatment of osteomyelitis has come to be the standard. After the treatment of the primary infection and stabilization of the fracture, physical therapy is started with the usual examinations and therapeutic program. The treatment is basically the same as that for fractures with normal healing.

After a long immobilization period, following unsuccessful surgical or nonsurgical treatment attempts, the electrolyte balance becomes deranged. This must be taken into account by the therapist, with the treatment procedures applied carefully and at a slowly increasing pace.

Nonsurgical Treatment

The examination results determine how the physical therapy should be applied. During the acute phase, the therapist usually follows the previously described guidelines for prophylactic treatment. However, to avoid additional irritation of the inflamed tissue, the stabilization of the injury must not be disturbed by these activities or manipulations.

After the infection has subsided, the treatment proceeds according to the usual schedule, modified for the individual symptoms and needs of the patient. The workload should never be excessive. The best dosage is determined by regular and intensive communication between the therapist and the physician.

ISCHEMIC CONTRACTURES

Muscle contractures due to insufficient blood supply are observed mainly after fractures close to the elbow, especially in children, and after fractures of the lower leg.

Mechanism of Injury

The insufficient blood supply of the muscles leads to cell death and to a permanent scarring and contracture of the affected muscle groups. The primary ischemia is caused by the compression of arteries by protruding bony edges of the proximal fragment in supracondylar fractures of the humerus (mechanical irritation of the brachial artery); by constrictive casts or bandages; by extensive hematomas; by edema; or by foreign material, such as surgical implants. Vascular spasms, overreaction of the sympathetic nerves, and venous congestion due to blocked drainage are also considered to be contributing factors. The decreased blood circulation results in edema and renewed pressure.

Types

The necrotic muscle tissue is replaced by scar tissue, which causes irreversible contractures of fingers and toes.

Clinical Considerations and Diagnosis

The acute signs are pain, loss of arterial pulse, decreased skin temperature, extensor loss of fingers and toes, local swelling, sensitivity to palpation, and disturbed sensory function.

The fully developed Volkmann's contracture is a flexion contracture of the wrist and the finger joints with atrophy of the flexors of the forearm. The metacarpophalangeal joints are hyperextended resulting in a clawhand with the thumb in abduction. Frequently, there is a simultaneous lesion of the median nerve, which results in a loss of the sensory function of the hand.

Medical Management

The treatment is directed at restitution of the surrounding arterial blood supply (cast removal, open reduction, incision of the muscle fascia). The fully developed contracture requires careful exercise therapy. Corrective measures by splints are frequently indicated. If these treatments are not successful, a surgical correction is indicated, including repositioning of the tendons, bone shortening, and neurolysis. Afterward, intensive physical and occupational training are necessary.

Physical Therapy

Nonsurgical Treatment

The therapeutic program should include the following procedures:

- improving the local blood supply
- mobilization: treatment of the contracture by active and active-assisted passive movements, supported by splints and slings
- muscle stimulation, possibly electrotherapy
- strengthening the muscles of the fingers, hand, and arm
- training in activities of daily living.

Surgical Treatment

After the immobilization period, the physical therapy includes the normal procedures: edema prophylaxis, improvement of the local blood supply, mobilization, strengthening exercises, and training in activities of daily living. Specific aspects and the intensity of the treatment are adjusted appropriately to the individual case at each treatment phase.

Bibliography

Aitken, A.P., H.K. Magill: Fractures involving the distal epiphyseal cartilage. J. Bone Jt Surg. 34–A (1952) 96

Blount, W.P.: Knochenbrüche bei Kindern. Thieme, Stuttgart 1975

Böhler, L.: Die Technik der Knochenbruchbehandlung. Maudrich, Wien 1957

Bruck-Gramcko, D.: Verletzungen der Hand. In Koslowski, L., Irmer, W., Bushe, K.; Lehrbuch der Chirurgie. Schattauer, Stuttgart 1978

Buck-Gramcko, D.: Sehnennähte. Im Dienste der Chirurgie, Heft 35. Ethicon, Hamburg 1979

Bunnell, S.: Surgery of the Hand. Lippincott, Philadelphia 1917

Burri, C., H. Ecke, E.H. Kuner, A. Pannike, L. Schweiberer, C.H. Schweikert, W. Spier, H. Tscherne: Unfallchirurgie. Springer, Berlin 1976

Colles, A.: On the fracture of the carpal extremity of the radius. Med. surg. J. 10 (1814) 182

Daniels, L.: Muskelfunktionsprüfung. Fischer, Stuttgart 1974

Debrunner, H.U.: AO Gelenkmessung (Neutral-0-Methode), Längenmessung, Umfangmessung. Dokumentation der BGOT, Tübingen 1971

Ehrenberg, H.: Über die Lösungs- und Atemtherapie von A. Schaarschuch Krankengymnastik 22 (1970) 176

Ehrenberg, H.: Atemtherapie in der Krankengymnastik. Krankengymnastik 23 (1971) 239

Ehrenberg, H.: Atemtherapie in der Krankengymnastik. Sonderheft der Krankengymnastik, ZV Krankengymnastik, München 1975

Frisch, H.: Manuelle Therapie in der Krankengymnastik. Krankengymnastik (1976) 93

Gillert, O.: Niederfrequente Reizströme in der therapeutischen Praxis. Pflaum, München 1977

Gillert, O.: Hydrotherapie und Balneotherapie in Theorie und Praxis. Pflaum, München 1978

Goldthwait, J.E.: Permanent dislocation of the patella. Ann. Surg. 29 (1899) 62

Grober, J.: Klinisches Lehrbuch der Physikalischen Therapie. Fischer, Jena 1963

Heipertz, W.: Sportmedizin, 6. Aufl. Thieme, Stuttgart 1980

Honigmann, M.: Krankengymnastische Behandlungsschwerpunkte nach traumatischen Sehnen- und Nervenverletzungen. Krankengymnastik (1977) 208

Iselin, M.: Chirurgie der Hand. Thieme, Stuttgart 1965

Jentschura, G., H.W. Janz: Beschäftigungstherapie, Grundlagen und Praxis, 2 Bde. Thieme, Stuttgart 1979

Kaltenborn, F.: Manuelle Therapie der Extremitätengelenke. DGMT, Hamm 1973

Klein-Vogelbach, S.: Funktionelle Bewe-gungslehre. Springer, Berlin 1976

Knott, M.: Komplexbewegungen. Fischer, Stuttgart 1970

Koppelmann, J.: Zur konservativen Behandlung und Anwendung der Kryotherapie bei postraumatischen und postoperativen Gelenkkontrakturen. Mschr. Unfallheilk. 74 (1971) 544

Koppelmann, J.: Eisanwendung als Ergänzung zur krankengymnastischen Behandlung. Oskar-Helene-Heim, FU Berlin 1974

Küntscher, G.: Praxis der Marknagelung. Schattauer, Stuttgart 1962

Lewit, K.: Manuelle Medizin. Barth, Leipzig 1977

List, M.: Die krankengymnastische Behandlung der verletzten Hand. Krankengymnastik (1968) 409

List, M.: Eisbehandlung in der Krankengymnastik. ZV Krankengymnastik, München 1978

List, M.: Die krankengymnastische Behandlung in der Traumatologie. Springer, Berlin 1978

Lorenzen, H.: Lehrbuch des Versehrtensports. Enke, Stuttgart 1961

Mittelbach, H.R.: Die verletzte Hand. Springer, Berlin 1977

Moberg, E.: Dringliche Handchirurgie, 3. Aufl. Thieme, Stuttgart 1972

von Mühlmann, A.: Krankengymnastik bei Verletzungsfolgen am Bewegungsapparat, 5. Aufl. Pflaum, München 1975

Müller, M.E., M. Allgöwer, R. Schneider, H. Willenegger: Manual der Osteosynthese, AO-Technik. Springer, Berlin 1969

Mumenthaler, M., H. Schliack: Läsionen peripherer Nerven, 4. Aufl. Thieme, Stuttgart 1982

Pauwels, F.: Atlas zur Biomechanik der gesunden und kranken Hüfte. Springer, Berlin 1973

Perren, S.M., A. Huggler, M. Russenberger, M. Allgöwer, R. Mathys, R. Schenk, H. Willenegger, M.E. Müller: The reaction of cortical bone to compression. Acta orthop. scand., Suppl. 125 (1969) 19–28

Petracic, B.: Funktionelle Nachbehandlung operierter Knochenbrüche, 2. Aufl. Thieme, Stuttgart 1983

Poelchen, R.: Selbstinnervationsbehandlung geschlossener Knochenbrüche und Verrenkungen, eine biologische Behandlungsart. Hippokrates, Stuttgart 1940

Ricklin, P., A. Rüttimann, M.S. del Buono: Die Meniskusläsion, 2. Aufl. Thieme, Stuttgart 1980

Schlosser, V., E. Kuner: Traumatologie, 3. Aufl. Thieme, Stuttgart 1980

Specht, G.: Primäre funktionelle. Behandlung nach Oberarmschaftbrüchen. Akt. Chir. 11 (1976) 227

Specht, G., O. Scheibe, R. Kreft: Ergebnisse der primären funktionellen Behandlung von Oberarmschaftbrüchen. Akt. Chir. 14 (1979) 249

Stenger, E.: Verbandlehre. Urban & Schwarzenberg, München 1969

Zichner, L.: Electrical stimulation in the treatment of pseudarthrosis. In: pseudarthroses and Their Treatment. 8th International Symposium on Typical Problems in Orthopedic Surgery. Thieme, Stuttgart 1979

Index

A

B

C

I

Q

R

S

T

About the Authors

ULRIKE BERGMANN was educated as a physical therapist at the University of Hamburg, receiving her degree there in 1975. From that time until 1980 she worked at a first-aid clinic in Boberg, specializing in traumatology. Since 1980 Ms. Bergmann has concentrated on physical therapy in the areas of pediatrics and neurology. In 1986, she began the practice of Vojta therapy.

WOLFGANG HEIPERTZ studied medicine at the universities of Tubingen, Vienna, and in Munich, and received his medical degree from the University of Berlin. In 1962 he qualifed as a full professor at the University of Heidelberg. Dr. Heipertz specializes in pediatric orthopedics, work-related injuries, sports medicine and sports injuries, physiotherapy, diseases of the spine, and artificial joint replacement. He has been chief surgeon at the Emergency Hospital in Tubingen and is now professor and medical director of the Orthopedic Hospital at J.-W. Goethe University in Frankfurt/Main.

BARBARA MORGENROTH is a licensed physical therapist who completed her training at the Hessisch Lichtenau School for Physiotherapy in 1965. She participated in internships in Bad Oeynhausen and Berlin. Ms. Morgenroth worked in England on the rehabilitation unit of St. Benedict's Hospital, then specialized in the rehabilitation of lepers and polio and tuberculosis patients. She worked with the German Society to Assist with Development in the Third World (Deutsche Entwicklungs Dienst) in Uganda and in Cameroun. Since 1975 she has been a staff member at the Hamburg Trade Union Emergency Hospital and is also involved in the theoretical and practical training of student physical therapists at the same institution.

LUDWIG ZICHNER studied medicine at the universities of Tubingen, Freiburg, and Vienna. He received his M.D. degree from the University of Dusseldorf in 1966. He began specializing in orthopedics in 1972 and qualified as full professor at the J.-W. Goethe University in Frankfurt/Main in 1984. As a member of the U.S. Orthopedic Research Society, he spent one month a year for ten years (1974-1984) as a visiting professor at the University of Arkansas Department of Orthopedics and was a special fellow in the Cleveland Clinic's Department of Orthopedics. Professor Zichner's clinical work has been concerned with the pathology of the bones and bone healing as well as with the problems of securing joint transplants, and he has published widely on those topics. Currently, he is chief surgeon at, and deputy medical director of, the Orthopedic University Hospital Friedrichsheim/Frankfurt am Main.

About the Editors

MARY ANNE RINEHART is Director of Physical Therapy at the Santa Clara Valley Medical Center in San Jose, California. She received a B.S. in physical therapy from the University of Kansas in 1962 and was awarded an M.S. in physical therapy by the Medical College of Virginia in 1980. Ms. Rinehart has practiced physical therapy for 19 years. Her special interests are improving methods of evaluating and treating neuropathological conditions and studying the effectiveness of physical therapy techniques.

TOM SUTTON is Director of Physical Therapy at Ralph K. Davies Medical Center in San Francisco, California. He received a bachelor of science degree in biology (1974) from New Mexico Tech, Socorro, N.M., and a certificate of physical therapy (1975) from the University of Pennsylvania. He recently completed a masters in business administration at the University of Phoenix, San Jose campus. Mr. Sutton has been a practicing physical therapist for 11 years.